001592
001590ୃ

WITHDRAWN

D1428437

cleu

o/3

581.9/CLA

FLORA OF THE BRITISH ISLES

ILLUSTRATIONS

ROTHAMSTED
EXP. STATION

HARPENDEN

A. R. CLAPHAM   T. G. TUTIN   E. F. WARBURG

FLORA OF THE BRITISH ISLES

ILLUSTRATIONS

PART III

BORAGINACEAE–COMPOSITAE

DRAWINGS BY

SYBIL J. ROLES

ROTHAMSTED
EXP. STATION
WITHDRAWN
1 5 APR 1986
HARPENDEN

OGG (WEST) BUILDING
LIBRARY
WITHDRAWN

CAMBRIDGE

AT THE UNIVERSITY PRESS

1963

Ogg

581.9/CLA

PUBLISHED BY

THE SYNDICS OF THE CAMBRIDGE UNIVERSITY PRESS

Bentley House, 200 Euston Road, London, N.W. 1
American Branch: 32 East 57th Street, New York 22, N.Y.
West African Office: P.O. Box 33, Ibadan, Nigeria

©

CAMBRIDGE UNIVERSITY PRESS

1963

Printed in Great Britain at the University Press, Cambridge
(Brooke Crutchley, University Printer)

PREFACE

The third part of these Illustrations of British plants follows the same general pattern as the first and second and completes the Dicotyledons.

The sequence followed and the nomenclature adopted are those used in *Flora of the British Isles*, second edition.

The drawings have, once again, almost all been made from fresh specimens and it is hoped that they will be useful in conjunction with either the *Flora* or the *Excursion Flora*.

We would again express our gratitude to Miss Roles, who has now been drawing plants for about fifteen years, for the skill and efficiency with which she has carried out the task.

We are greatly indebted to the following for supplying many of the plants illustrated in this Part: Miss J. Allison, P. W. Ball, Mrs J. M. Ball, C. H. Barnes, Miss M. S. Campbell, Mrs P. A. Candlish, Miss A. P. Conolly, Mrs G. Crompton, R. W. David, K. W. Dent, Miss U. K. Duncan, J. S. L. Gilmour, the late R. A. Graham, A. Hackett, G. Halliday, M. K. Hanson, C. B. Heginbotham, Miss J. E. Hibbard, E. K. Horwood, Mrs J. M. Horwood, J. H. Lacey, J. E. Lousley, D. McClintock, Miss E. M. Medwin, R. D'O. Good, C. D. Pigott, M. E. D. Poore, M. C. F. Proctor, C. E. Raven, J. E. Raven, F. Rose, Miss E. M. Rosser, P. D. Sell, G. M. Spooner, Mrs F. le Sueur, F. J. Taylor, Mrs L. M. Taylor, S. M. Walters, D. A. Webb, Mrs E. W. Woodward, and P. F. Yeo.

A. R. C.
T. G .T.
E. F. W.

PART III

BORAGINACEAE–COMPOSITAE

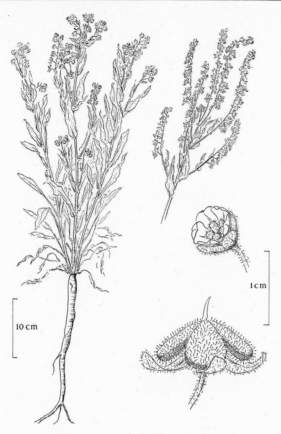

1013. *Cynoglossum officinale* L. Hound's-tongue Purplish

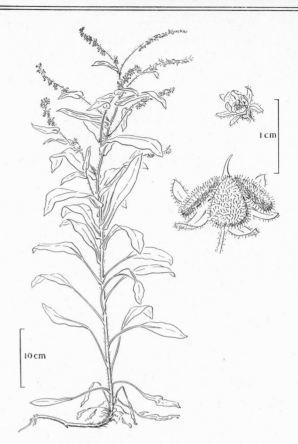

1014. *Cynoglossum germanicum* Jacq. Green Hound's-
tongue Purplish

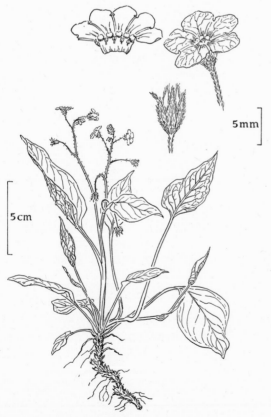

1015. *Omphalodes verna* Moench Blue-eyed Mary Blue

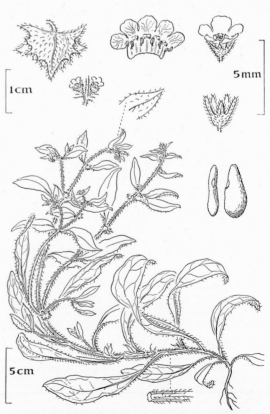

1016. *Asperugo procumbens* L. Madwort Purplish,
becoming blue

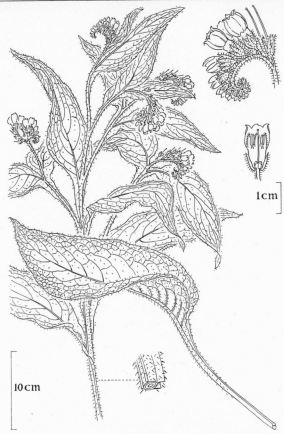

1017. *Symphytum officinale* L.   Comfrey   Whitish
or purplish

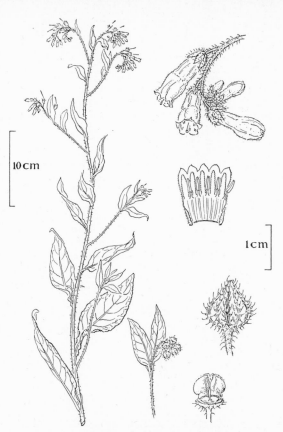

1018. *Symphytum asperum* Lepech.   Rough Comfrey
Pink, becoming blue

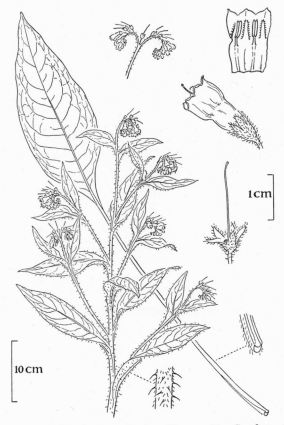

1019. *Symphytum × uplandicum* Nyman   Blue Comfrey
Blue or purplish

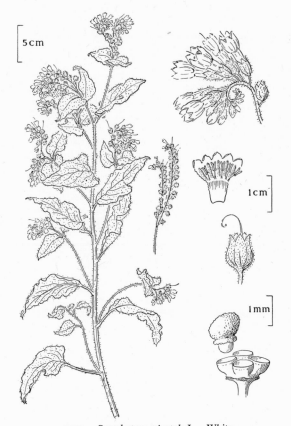

1020. *Symphytum orientale* L.   White

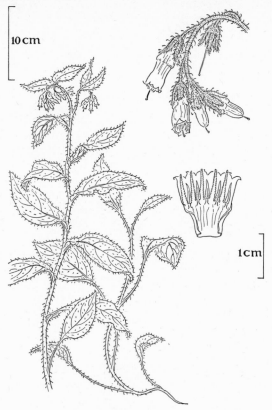

1021. *Symphytum tuberosum* L. Tuberous Comfrey Yellowish-white

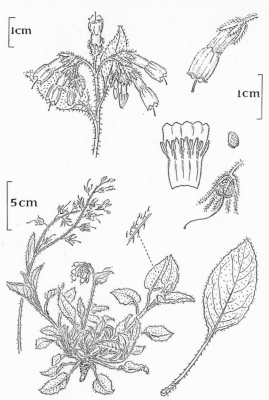

1022. *Symphytum grandiflorum* DC. Yellowish-white

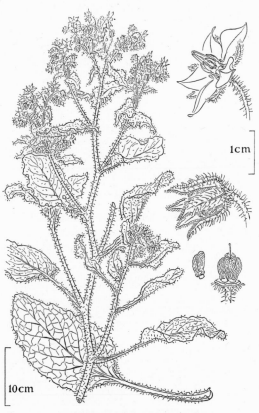

1023. *Borago officinalis* L. Borage Blue

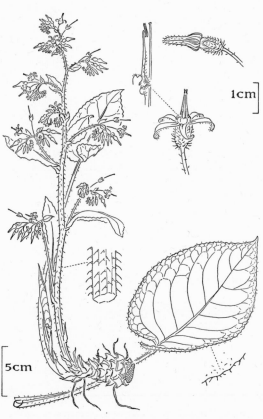

1024. *Trachystemon orientalis* (L.) G. Don Blue

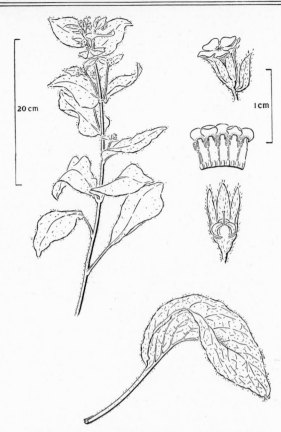

1025. *Pentaglottis sempervirens* (L.) Tausch   Alkanet
Blue

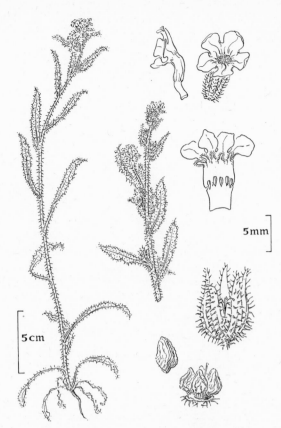

1026. *Anchusa arvensis* (L.) Bieb.   Bugloss   Blue

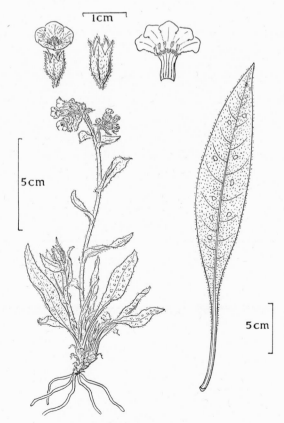

1027. *Pulmonaria longifolia* (Bast.) Bor.   Lungwort
Pink, becoming blue

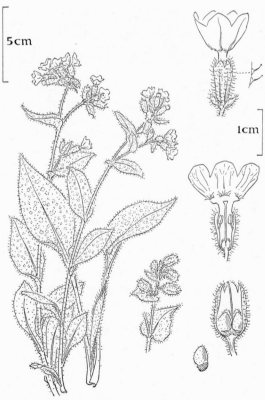

1028. *Pulmonaria officinalis* L.   Lungwort   Pink,
becoming blue

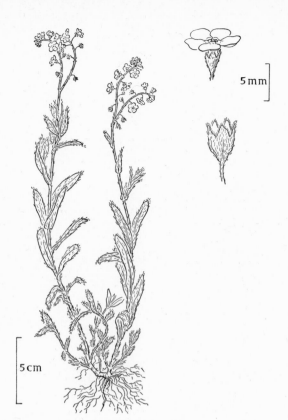

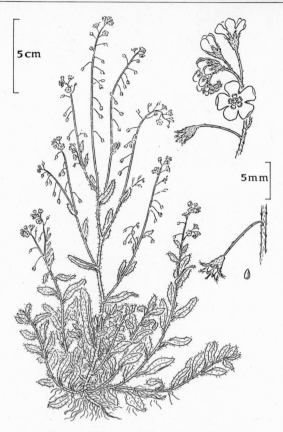

1029.   *Myosotis scorpioides* L.   Water Forget-me-not
        Blue

1030.   *Myosotis secunda* A. Murr.   Water Forget-me-not
        Blue

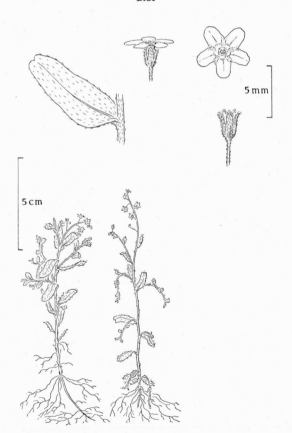

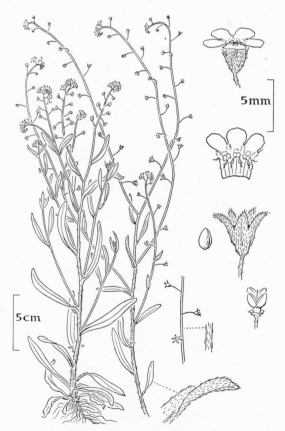

1031.   *Myosotis brevifolia* C. E. Salmon   Pale blue

1032.   *Myosotis caespitosa* K. F. Schultz   Water
        Forget-me-not   Blue

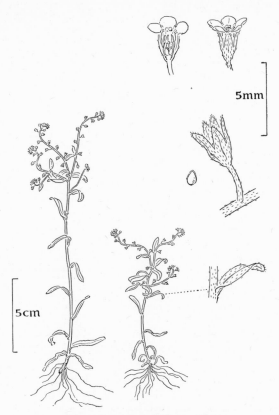

1033. *Myosotis sicula* Guss. Blue

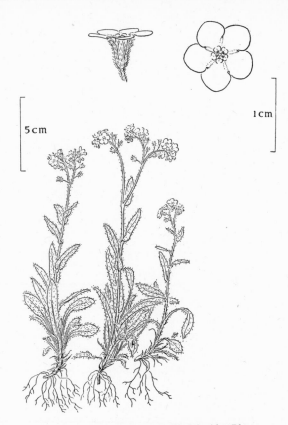

1034. *Myosotis alpestris* F. W. Schmidt Blue

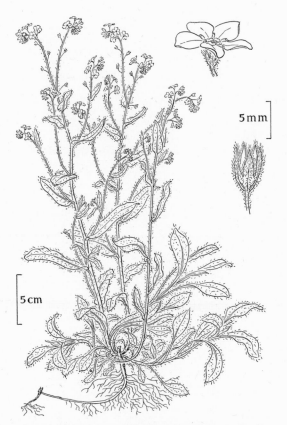

1035. *Myosotis sylvatica* Hoffm Wood Forget-me-not
Blue

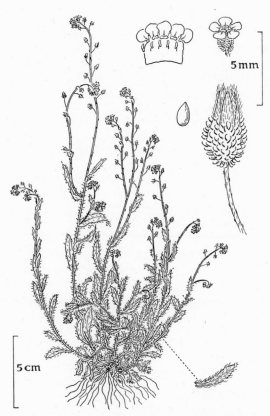

1036. *Myosotis arvensis* (L.) Hill Common Forget-me-not
Blue

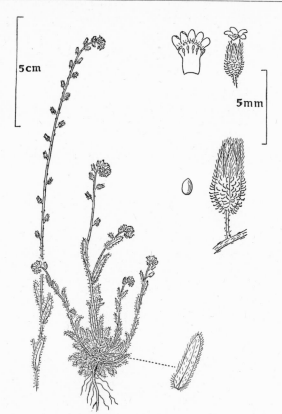

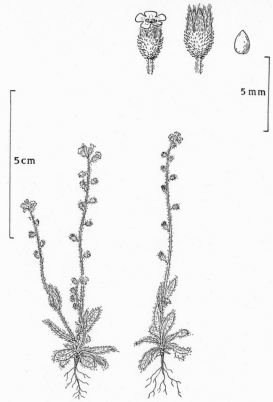

1037. *Myosotis discolor* Pers.   Yellow and blue Forget-
        me-not   Yellow, becoming blue

1038. *Myosotis ramosissima* Rochel   Early Forget-me-not
        Blue

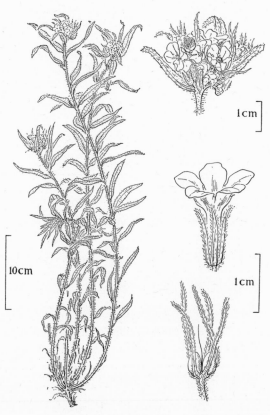

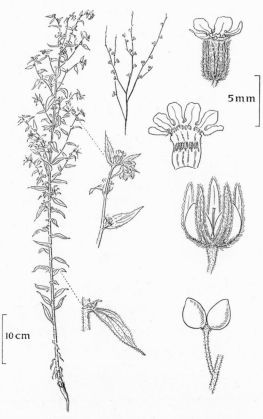

1039. *Lithospermum purpurocaeruleum* L.   Blue Gromwell
        Reddish-purple, becoming blue

1040. *Lithospermum officinale* L.   Gromwell   White

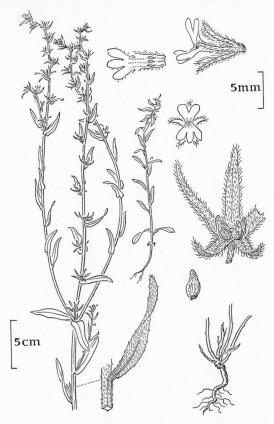

1041. *Lithospermum arvense* L.   Corn Gromwell   White

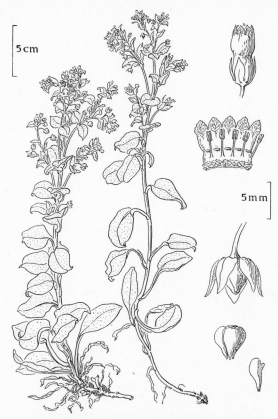

1042. *Mertensia maritima* (L.) S. F. Gray   Northern Shore-wort   Pink, becoming blue and pink

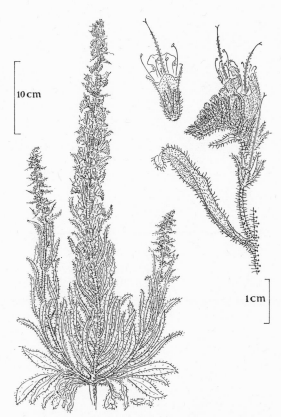

1043. *Echium vulgare* L.   Viper's Bugloss   Blue

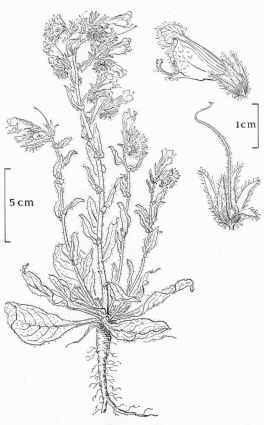

1044. *Echium lycopsis* L.   Purple Viper's Bugloss   Red, becoming purple-blue

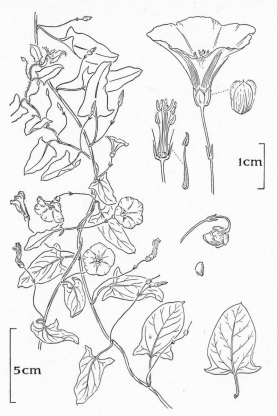

1045. *Convolvulus arvensis* L.   Bindweed   White or pink

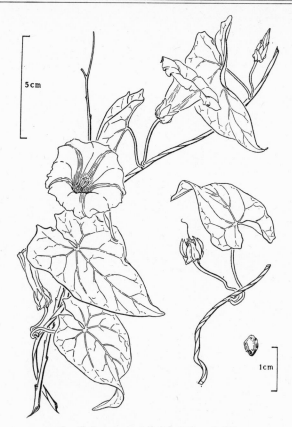

1046. *Calystegia sepium* (L.) R.Br.   Bellbine
White or pink

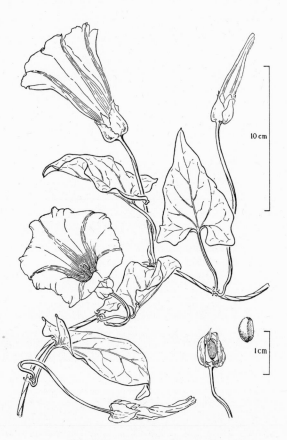

1047. *Calystegia sepium* ssp. *silvatica* (Kit.) Maire
White

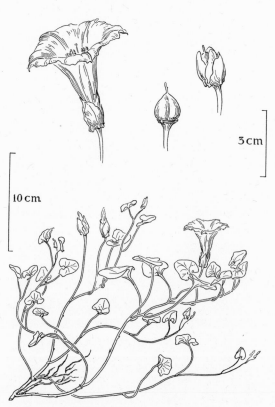

1048. *Calystegia soldanella* (L.) R.Br.   Sea Bindweed
Pink or pale purple

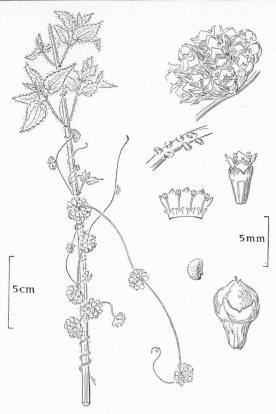

1049. *Cuscuta europaea* L.   Large Dodder
Pinkish-white

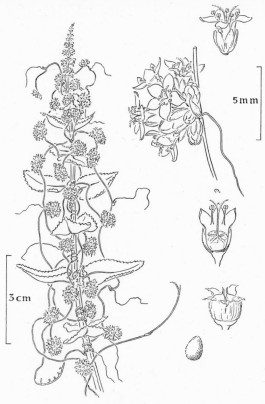

1050. *Cuscuta epithymum* (L.) L.   Common Dodder
Pinkish

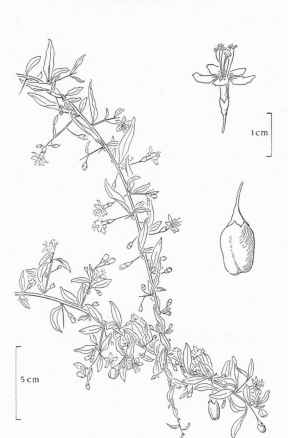

1051. *Lycium halimifolium* Mill.   Duke of Argyll's
Tea-plant   Rose-purple becoming pale brown

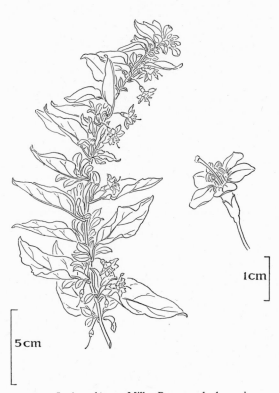

1052. *Lycium chinense* Mill.   Rose-purple, becoming
pale brown

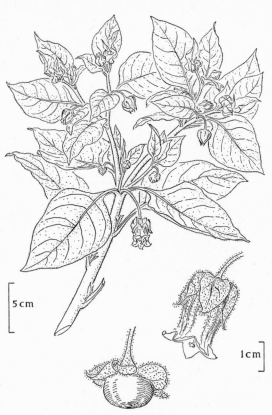

1053. *Atropa bella-donna* L.   Dwale, Deadly Nightshade
Violet or greenish

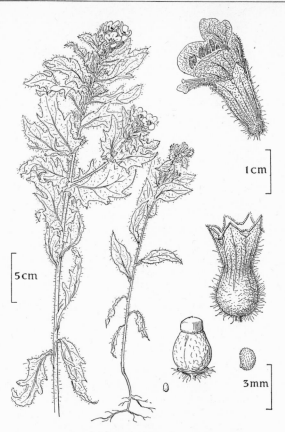

1054. *Hyoscyamus niger* L.   Henbane
Yellow, purple-veined

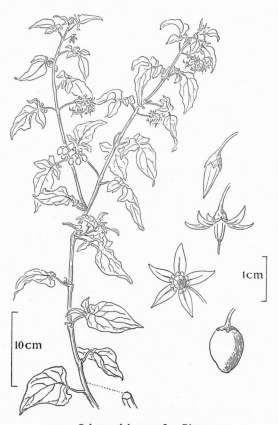

1055. *Solanum dulcamara* L.   Bittersweet,
Woody Nightshade   Purple

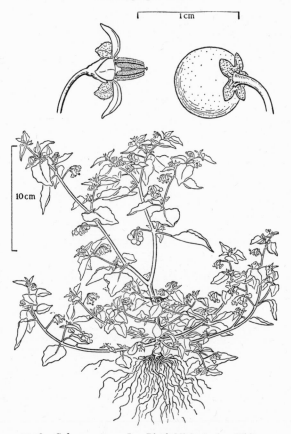

1056. *Solanum nigrum* L.   Black Nightshade   White

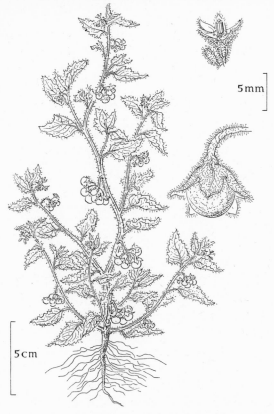

1057. *Solanum sarrachoides* Sendt. White

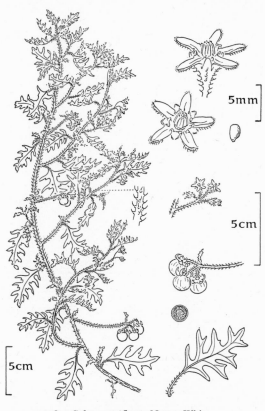

1058. *Solanum triflorum* Nutt. White

1059. *Datura stramonium* L. Thorn-apple
White or purple

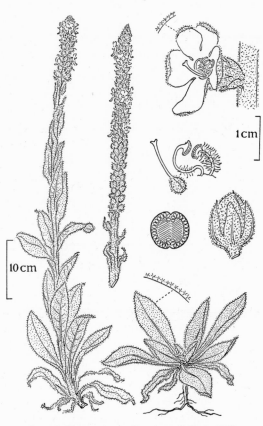

1060. *Verbascum thapsus* L. Aaron's Rod Yellow

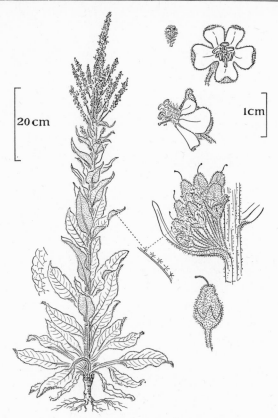

1061. *Verbascum lychnitis* L.   White Mullein
White or yellow

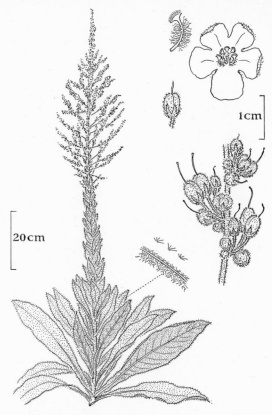

1062. *Verbascum pulverulentum* Vill.   Hoary Mullein
Yellow

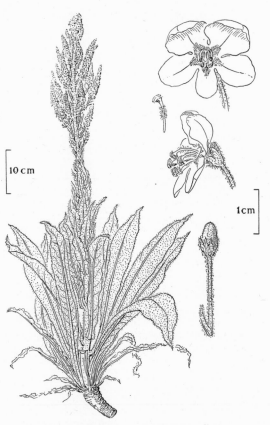

1063. *Verbascum speciosum* Schrad.   Yellow

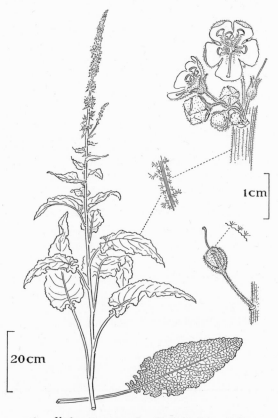

1064. *Verbascum nigrum* L.   Dark Mullein   Yellow

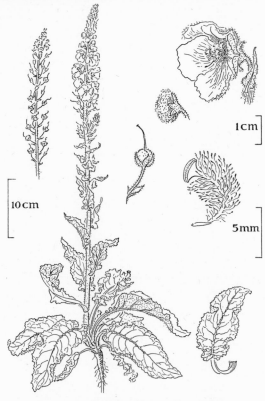

1065.   *Verbascum blattaria* L.   Moth Mullein
Yellow or whitish

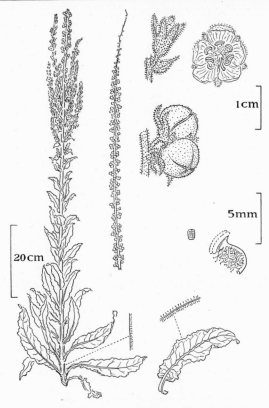

1066.   *Verbascum virgatum* Stokes   Twiggy Mullein
Yellow

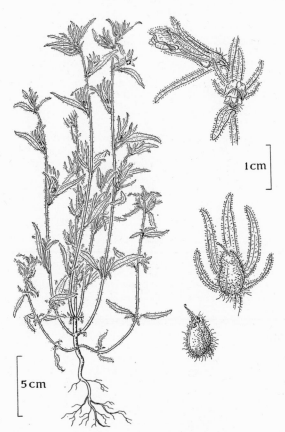

1067.   *Antirrhinum orontium* L.   Weasel's Snout, Calf's
Snout   Pinkish-purple

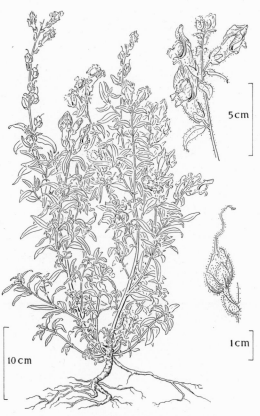

1068.   *Antirrhinum majus* L.   Snapdragon
Reddish-purple

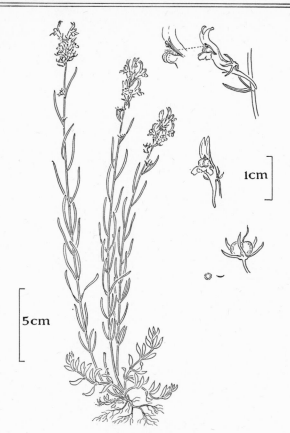

1069. *Linaria pelisseriana* (L.) Mill.
Violet and white

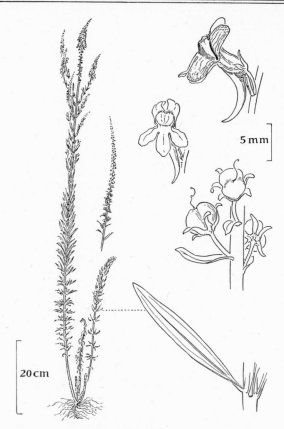

1070. *Linaria purpurea* (L.) Mill.   Purple Toadflax
Violet

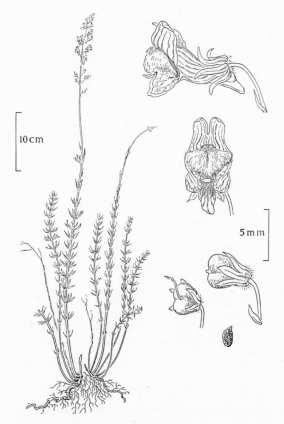

1071. *Linaria repens* (L.) Mill.   Pale Toadflax   White
or pale lilac, veined violet

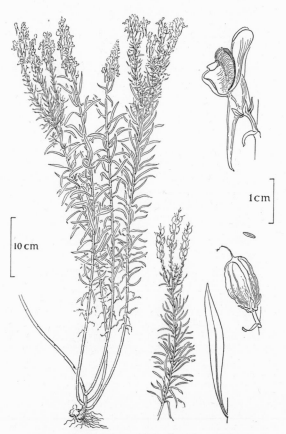

1072. *Linaria vulgaris* Mill.   Toadflax   Yellow

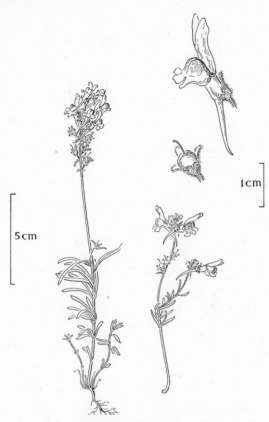

1073.   *Linaria supina* (L.) Chazelles   Yellow

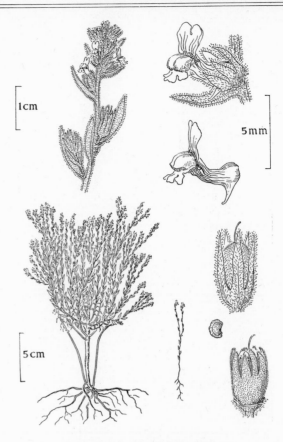

1074.   *Linaria arenaria* DC.   Yellow

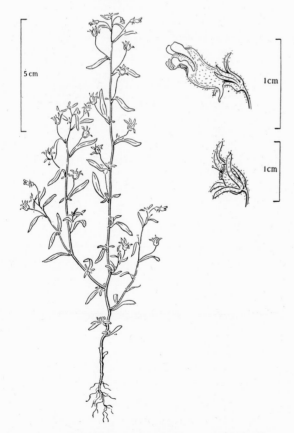

1075.   *Chaenorhinum minus* (L.) Lange   Small Toadflax
        Purplish

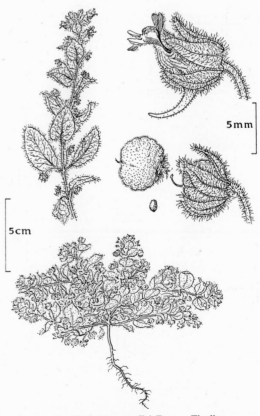

1076.   *Kickxia spuria* (L.) Dum.   Fluellen
        Yellow and purple

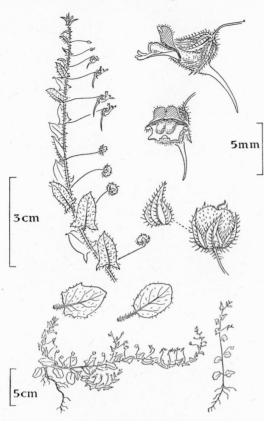

1077.   *Kickxia elatine* (L.) Dum.   Fluellen
Yellow and purple

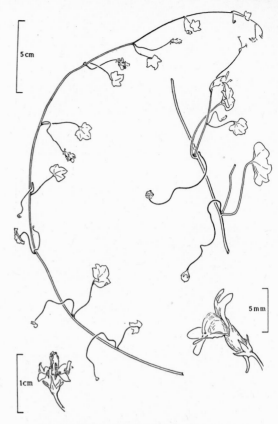

1078.   *Cymbalaria muralis* Gaertn., Mey. & Scherb.
Ivy-leaved Toadflax   Lilac and yellow

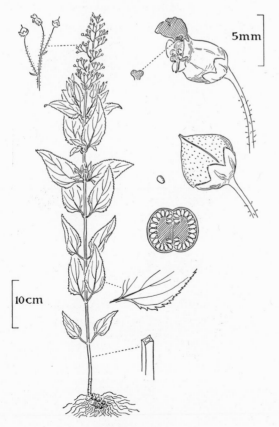

1079.   *Scrophularia nodosa* L.   Figwort
Reddish-brown

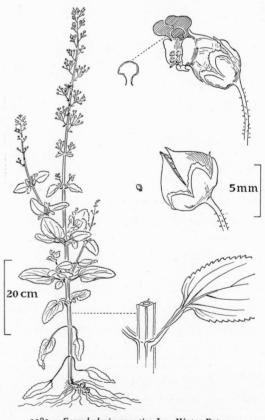

1080.   *Scrophularia aquatica* L.   Water Betony
Brownish-purple

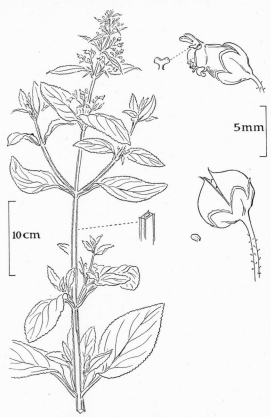

1081. *Scrophularia umbrosa* Dum.   Brownish-purple

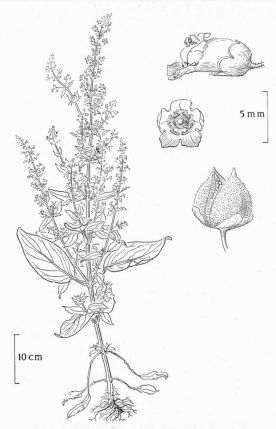

1082. *Scrophularia scorodonia* L.   Balm-leaved Figwort
Dull purple

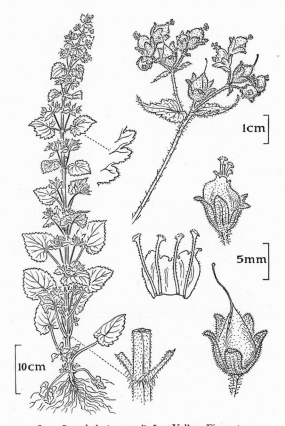

1083. *Scrophularia vernalis* L.   Yellow Figwort
Greenish-yellow

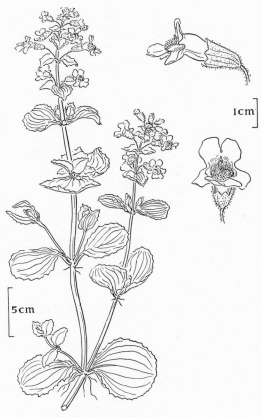

1084. *Mimulus guttatus* DC.   Monkey-flower
Yellow, red-spotted

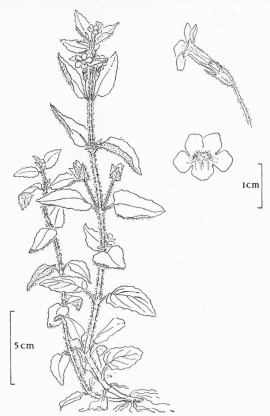

1085.   *Mimulus moschatus* Lindl.   Musk   Yellow

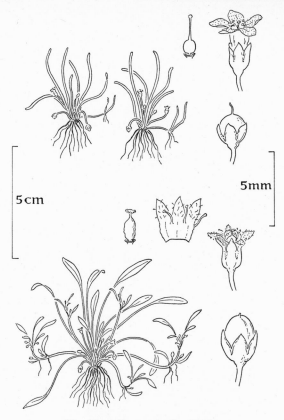

1086.   *Limosella aquatica* L.   Mudwort
White or lavender (bottom)
L. subulata Ives   White (top)

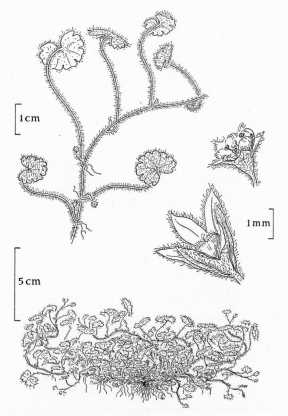

1087.   *Sibthorpia europaea* L.   Cornish Moneywort
Yellowish and pink

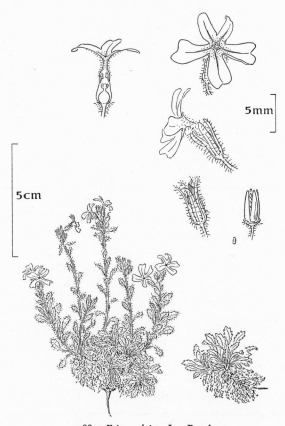

1088.   *Erinus alpinus* L.   Purple

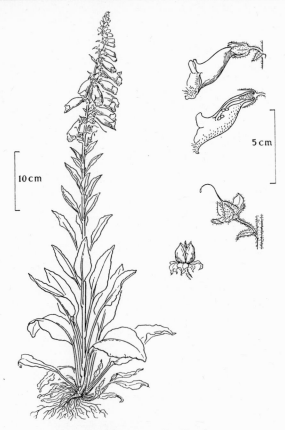

1089.   *Digitalis purpurea* L.   Foxglove
Pinkish-purple or white

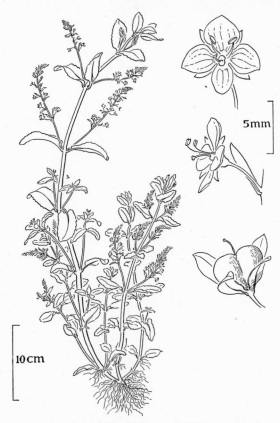

1090.   *Veronica beccabunga* L.   Brooklime   Blue

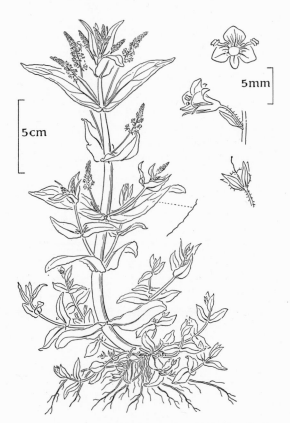

1091.   *Veronica anagallis-aquatica* L.   Water Speedwell
Pale blue

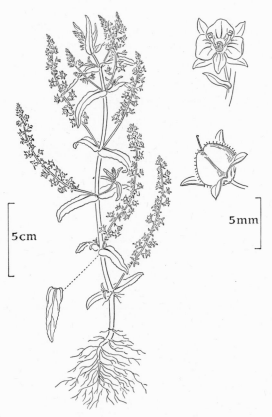

1092.   *Veronica catenata* Pennell   Pink

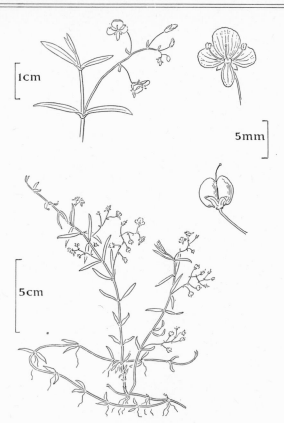

1093.   *Veronica scutellata* L.   Marsh Speedwell   White
        or pale blue

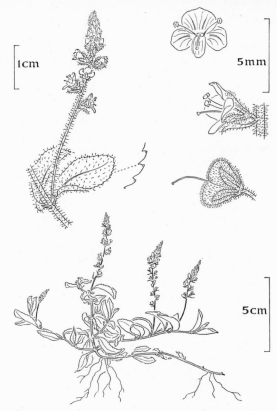

1094.   *Veronica officinalis* L.   Common Speedwell   Lilac

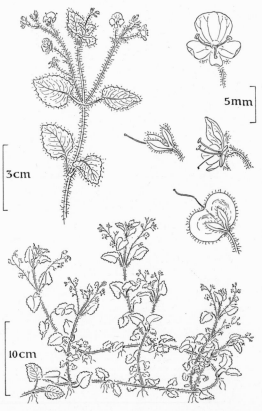

1095.   *Veronica montana* L.   Wood Speedwell   Lilac-blue

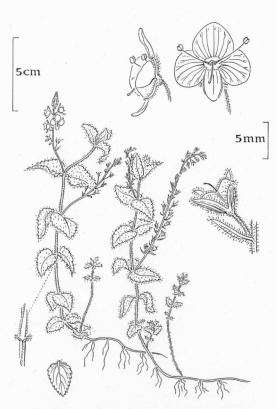

1096.   *Veronica chamaedrys* L.   Germander Speedwell
        Blue

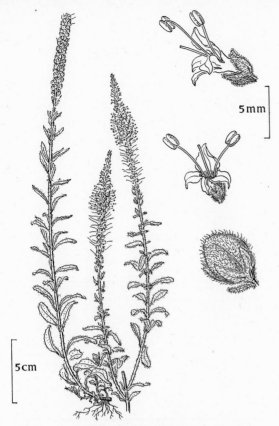

1097. *Veronica spicata* ssp. *hybrida* (L.) E. F. Warb.
       Spiked Speedwell   Blue

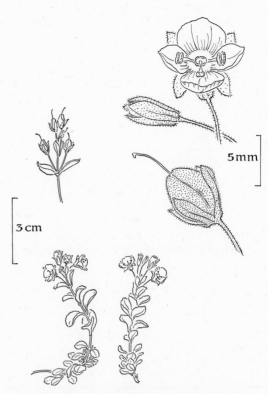

1098. *Veronica fruticans* Jacq.   Rock Speedwell   Blue

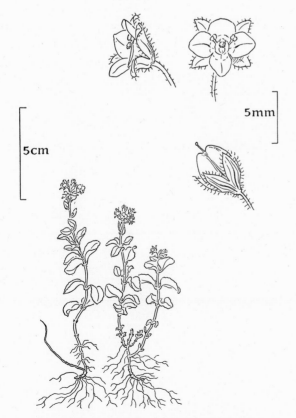

1099. *Veronica alpina* L.   Alpine Speedwell   Blue

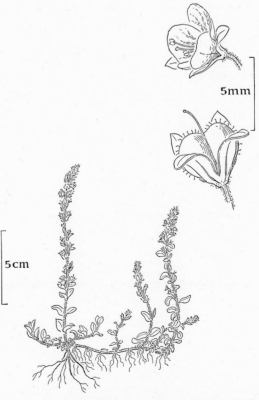

1100. *Veronica serpyllifolia* L.   Thyme-leaved Speedwell
       White or pale blue

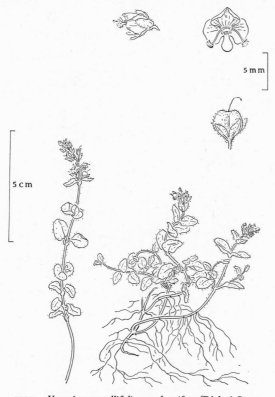

1101. *Veronica serpyllifolia* ssp. *humifusa* (Dicks.) Syme
Blue

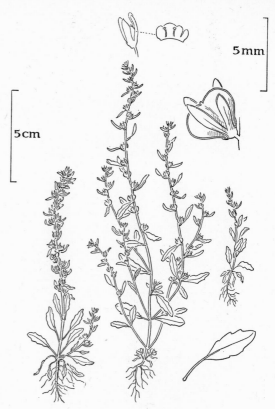

1102.   *Veronica peregrina* L.   Blue

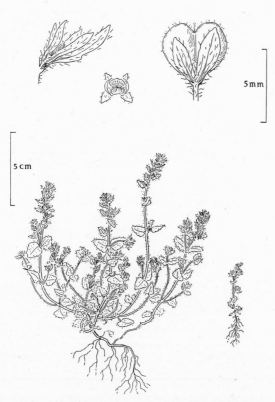

1103.   *Veronica arvensis* L.   Wall Speedwell   Blue

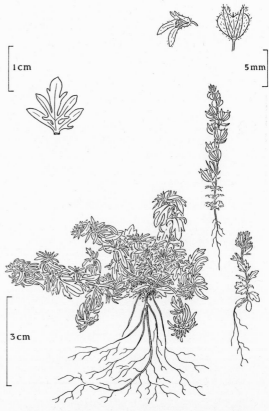

1104.   *Veronica verna* L.   Spring Speedwell   Blue

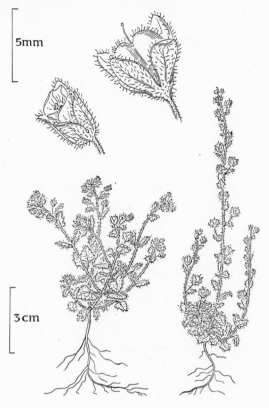

1105.   *Veronica praecox* All.   Blue

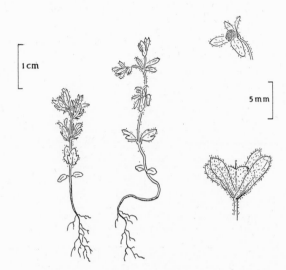

1106.   *Veronica triphyllos* L.   Fingered Speedwell   Blue

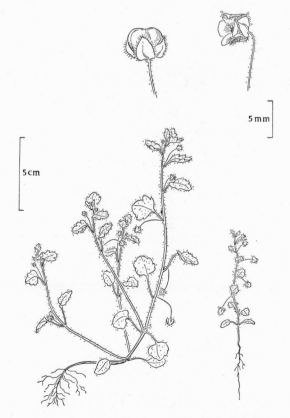

1107.   *Veronica hederifolia* L.   Ivy Speedwell   Pale lilac

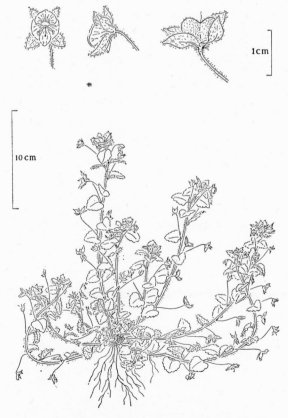

1108.   *Veronica persica* Poir.   Buxbaum's Speedwell   Blue

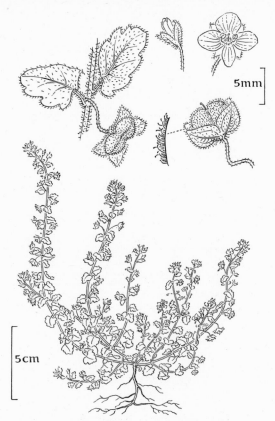

1109. *Veronica polita* Fr.   Grey Speedwell   Blue

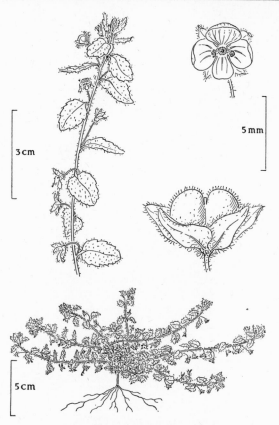

1110. *Veronica agrestis* L.   Field Speedwell
Bluish, white or pinkish

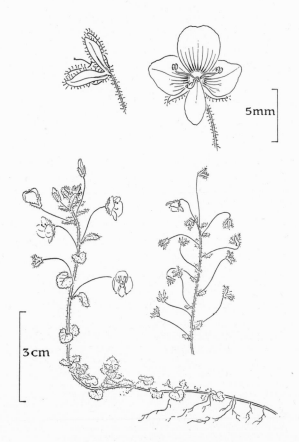

1111. *Veronica filiformis* Sm.   Blue

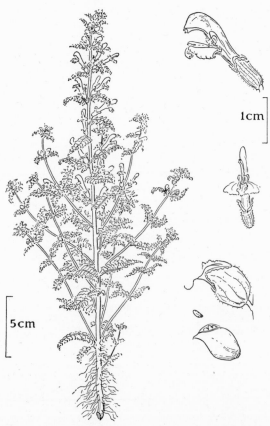

1112. *Pedicularis palustris* L.   Red-rattle   Purplish-pink

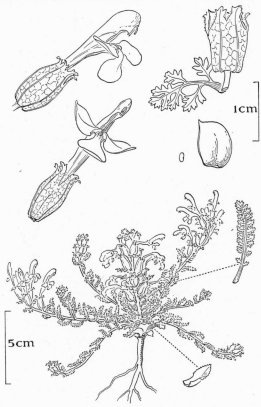

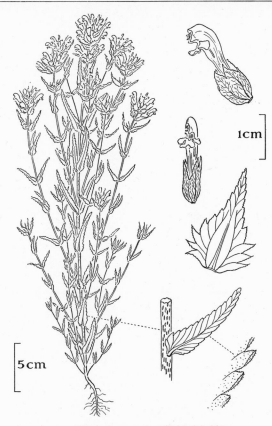

1113. *Pedicularis sylvatica* L. Lousewort Pink

1114. *Rhinanthus serotinus* (Schönh.) Oborny
Greater Yellow-rattle Yellow

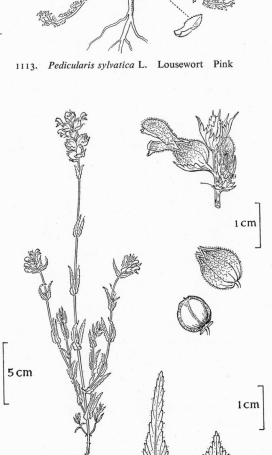

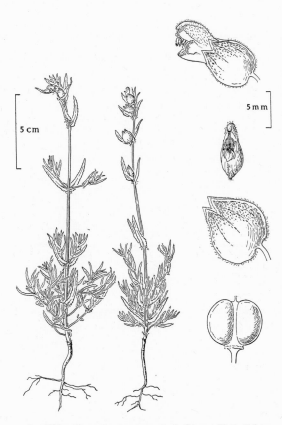

1115. *Rhinanthus minor* ssp. *stenophyllus* (Schur) O. Schwarz
Yellow-rattle Yellow

1116. *Rhinanthus minor* ssp. *monticola* (Sterneck) O. Schwarz
Yellow, becoming brown

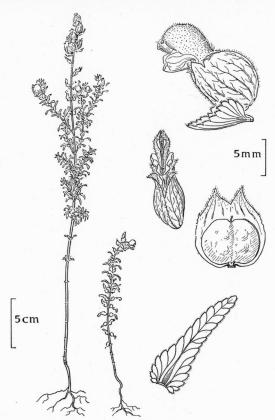

1117.   *Rhinanthus minor* ssp. *calcareus* (Wilmott) E. F. Warb.
Yellow

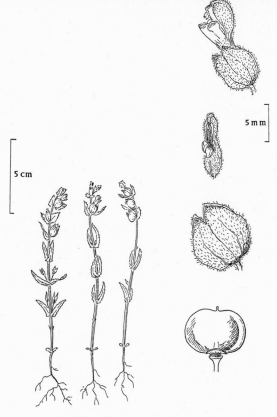

1118.   *Rhinanthus* × *gardineri* Druce   Yellow

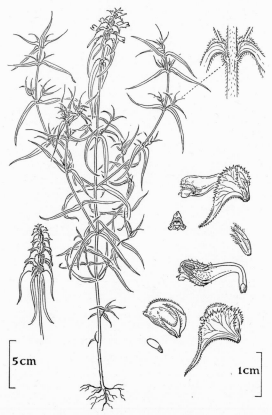

1119.   *Melampyrum cristatum* L.   Crested Cow-wheat
Yellow and purple

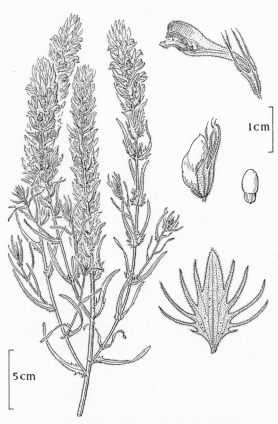

1120.   *Melampyrum arvense* L.   Field Cow-wheat
Pink

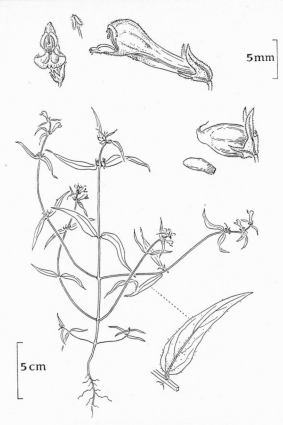

1121. *Melampyrum pratense* L.   Common Cow-wheat
Yellow to whitish

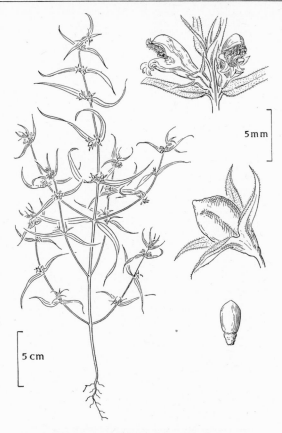

1122. *Melampyrum sylvaticum* L.   Wood Cow-wheat
Brownish-yellow

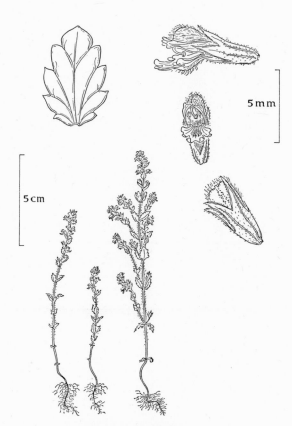

1123. *Euphrasia micrantha* Rchb.   Eyebright
White or purplish

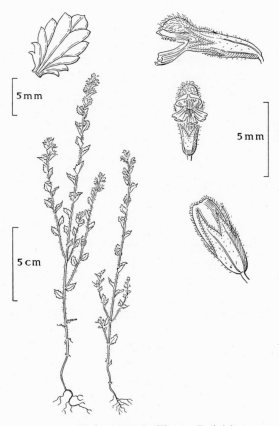

1124. *Euphrasia scottica* Wettst.   Eyebright
White or violet

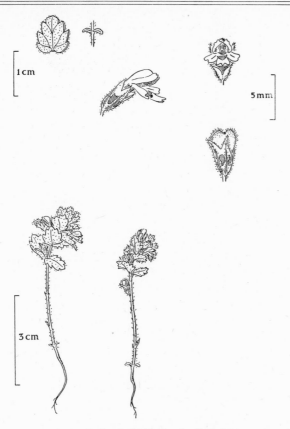

1125.   *Euphrasia frigida* Pugsl.   Eyebright
White and lilac

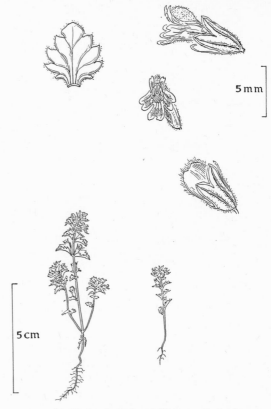

1126.   *Euphrasia foulaensis* Wettst.   Eyebright
Violet or white

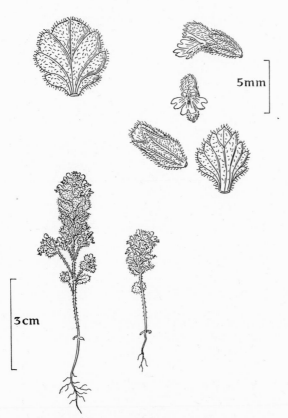

1127.   *Euphrasia rotundifolia* Pugsl.   Eyebright   White

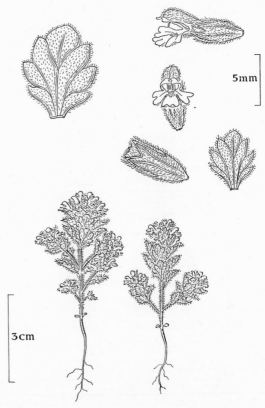

1128.   *Euphrasia marshallii* Pugsl.   Eyebright
White, rarely purplish

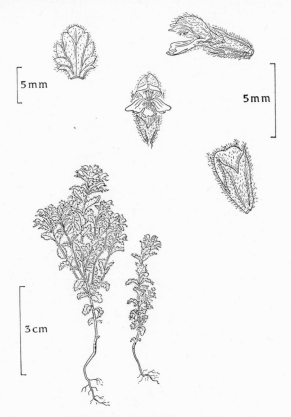

1129. *Euphrasia curta* (Fr.) Wettst. Eyebright White

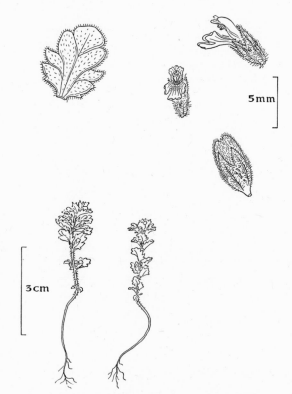

1130. *Euphrasia cambrica* Pugsl. Eyebright Whitish

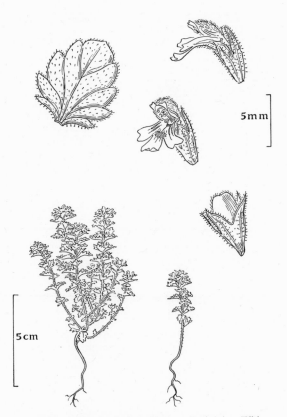

1131. *Euphrasia occidentalis* Wettst. Eyebright White

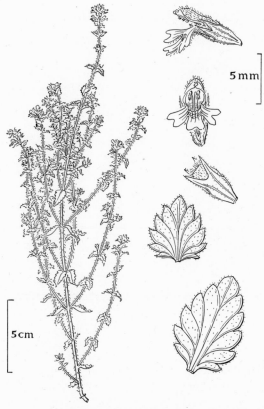

1132. *Euphrasia nemorosa* (Pers.) Wallr. Eyebright
White or bluish

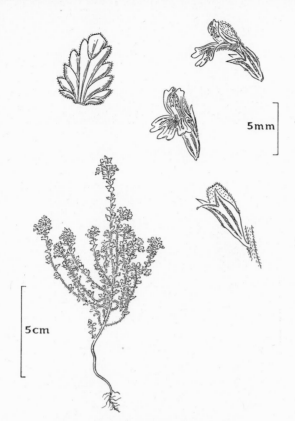

1133. *Euphrasia confusa* Pugsl.   Eyebright
White, purple or yellow

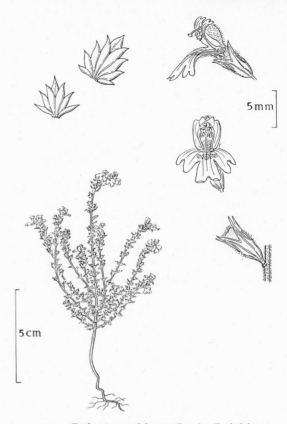

1134. *Euphrasia pseudokerneri* Pugsl.   Eyebright
White or bluish

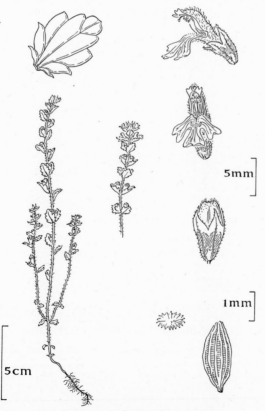

1135. *Euphrasia borealis* Wettst.   Eyebright
White or bluish

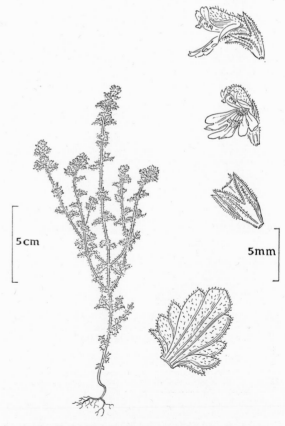

1136. *Euphrasia brevipila* Burnat & Gremli   Eyebright
Lilac or white and lilac

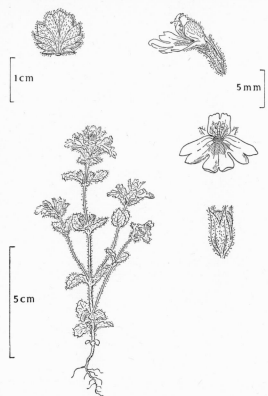

1cm

5mm

5cm

1137. *Euphrasia rostkoviana* Hayne Eyebright
White or lilac-tinted

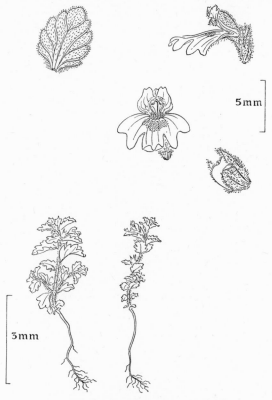

5mm

3mm

1138. *Euphrasia rivularis* Pugsl. Eyebright
White and purplish

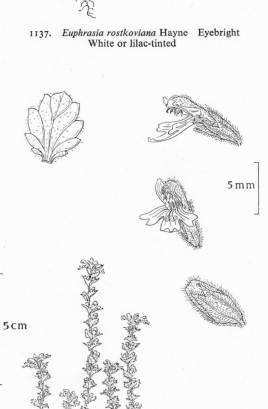

5mm

5cm

1139. *Euphrasia anglica* Pugsl. Eyebright
Whitish or lilac-tinted

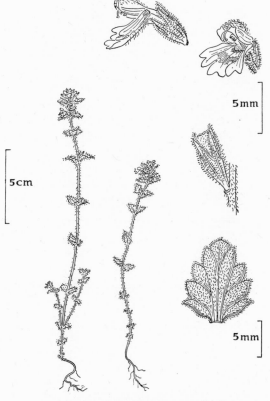

5mm

5cm

5mm

1140. *Euphrasia hirtella* Reut. Eyebright
White, bluish-tinted

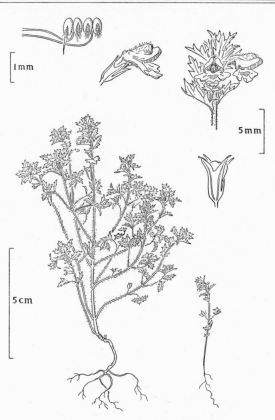

1141.   *Euphrasia salisburgensis* Funck   Eyebright
White

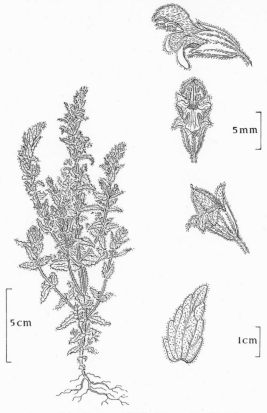

1142.   *Odontites verna* (Bell.) Dum.   Red Bartsia
Purplish-pink

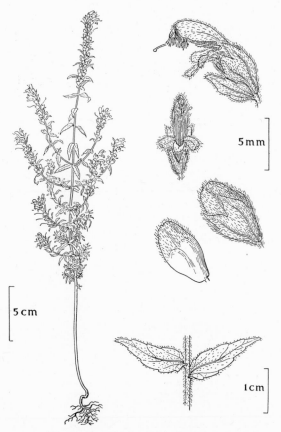

1143.   *Odontites verna* ssp. *serotina* (Wettst.) E. F. Warb.
Red Bartsia   Purplish-pink

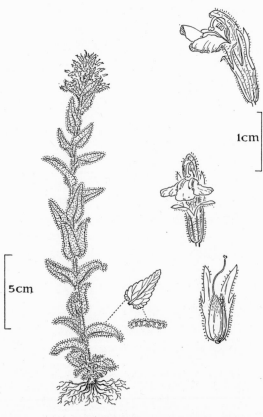

1144.   *Parentucellia viscosa* (L.) Caruel   Yellow Bartsia
Yellow

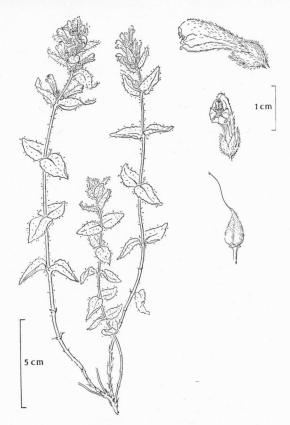

1145. *Bartsia alpina* L.   Alpine Bartsia   Purple

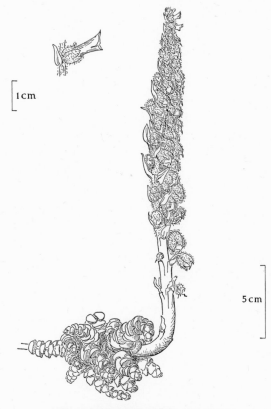

1146. *Lathraea squamaria* L.   Toothwort
White, purple-tinged

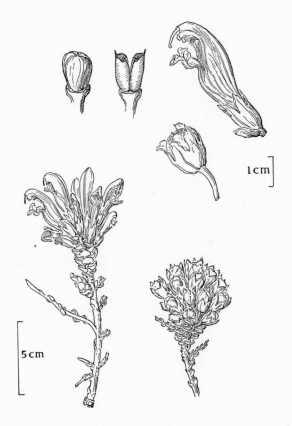

1147. *Lathraea clandestina* L.   Bright purple

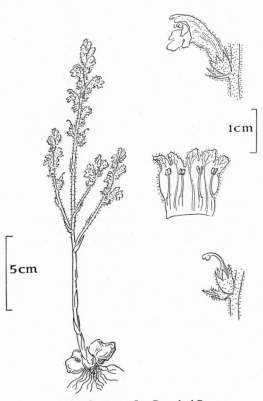

1148. *Orobanche ramosa* L.   Branched Broomrape
Yellowish-white

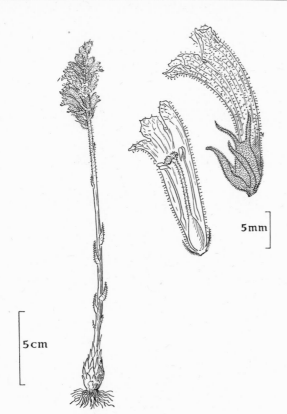

1149. *Orobanche purpurea* Jacq.   Purple Broomrape
Bluish-purple

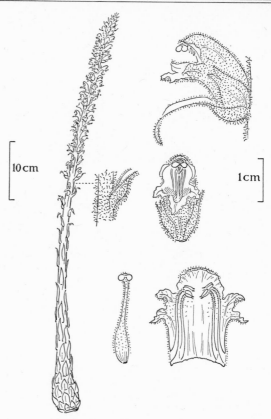

1150. *Orobanche rapum-genistae* Thuill.   Greater
Broomrape   Yellowish

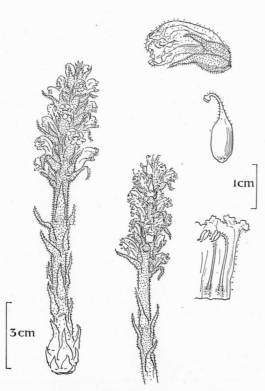

1151. *Orobanche alba* Willd.   Red Broomrape
Purplish-red

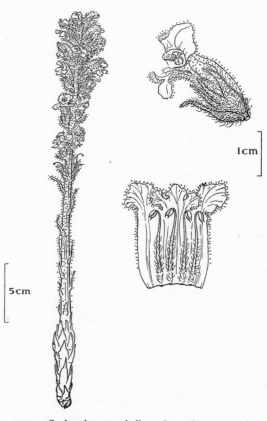

1152. *Orobanche caryophyllacea* Sm.   Clove-scented
Broomrape   Yellowish to purplish

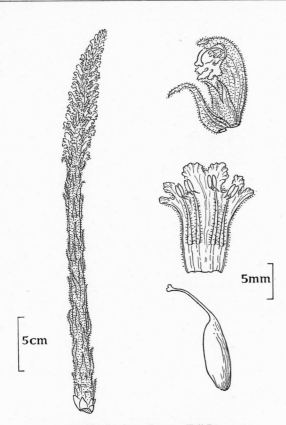

1153. *Orobanche elatior* Sutton   Tall Broomrape
Pale yellow, purple-tinged

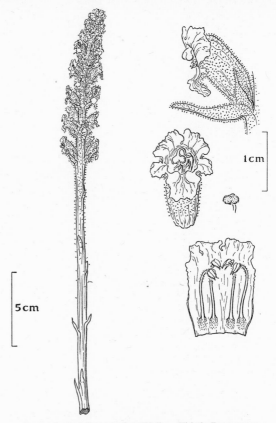

1154. *Orobanche reticulata* Wallr.   Thistle Broomrape
Yellowish, purple-tinged

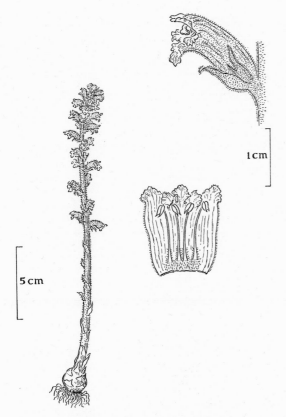

1155. *Orobanche minor* Sm.   Lesser Broomrape
Yellowish, purple-tinged

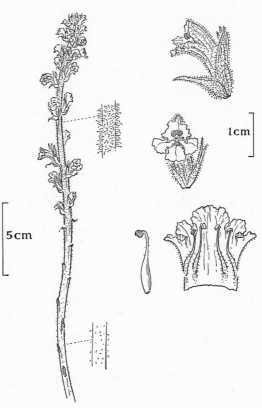

1156. *Orobanche picridis* Koch   Picris Broomrape
Yellowish, purple-tinged

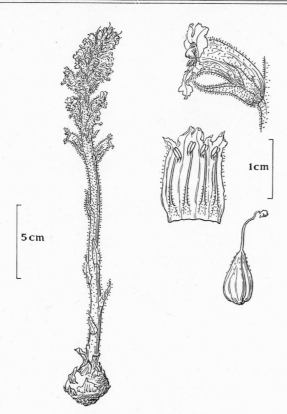

1157. *Orobanche hederae* Duby Ivy Broomrape
Cream, purple-veined

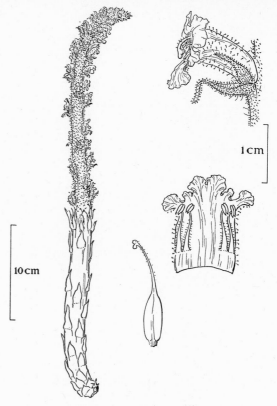

1158. *Orobanche maritima* Pugsl. Carrot Broomrape
Yellowish, purple-veined

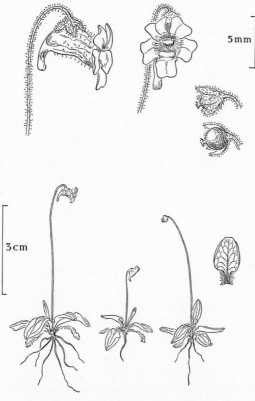

1159. *Pinguicula lusitanica* L. Pale Butterwort
Pale lilac

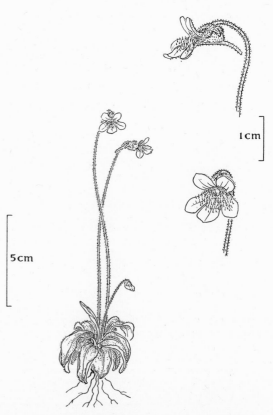

1160. *Pinguicula vulgaris* L. Common Butterwort
Violet and white

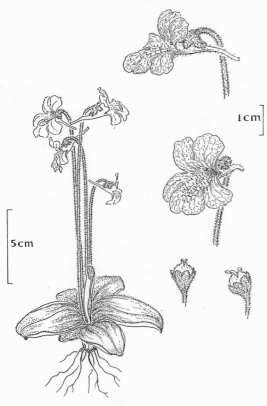

1161.   *Pinguicula grandiflora* Lam.   Violet and white

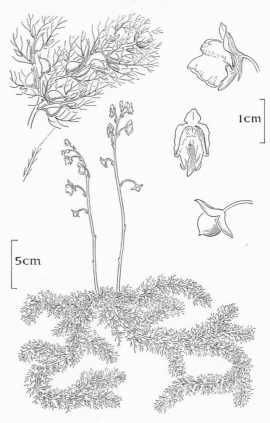

1162.   *Utricularia vulgaris* L.   Greater Bladderwort
Yellow

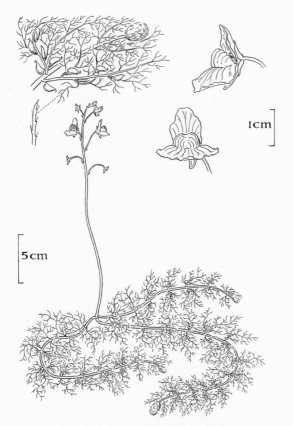

1163.   *Utricularia neglecta* Lehm.   Yellow

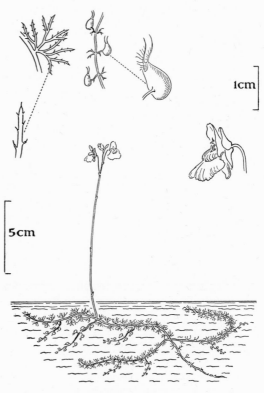

1164.   *Utricularia intermedia* Hayne   Intermediate
Bladderwort   Yellow

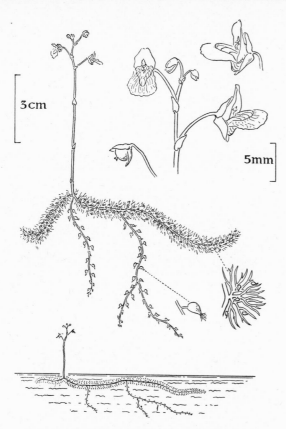

1165.   *Utricularia minor* L.   Lesser Bladderwort
Yellow

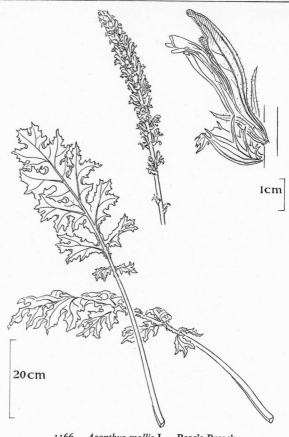

1166.   *Acanthus mollis* L.   Bear's Breech
Whitish, purple-veined

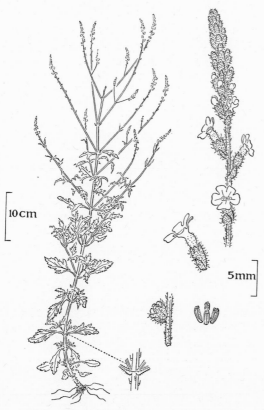

1167.   *Verbena officinalis* L.   Vervain   Pale lilac

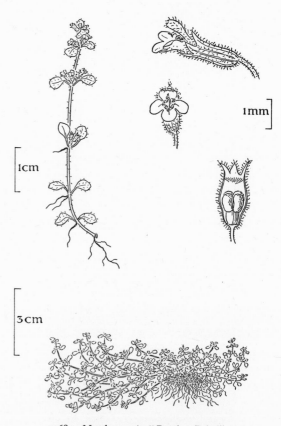

1168.   *Mentha requienii* Benth.   Pale lilac

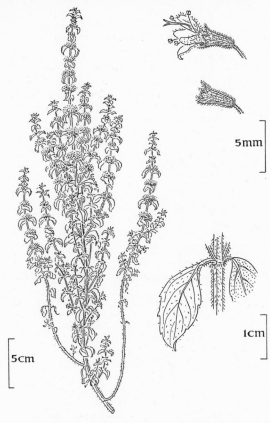

1169. *Mentha pulegium* L. Penny-royal Lilac

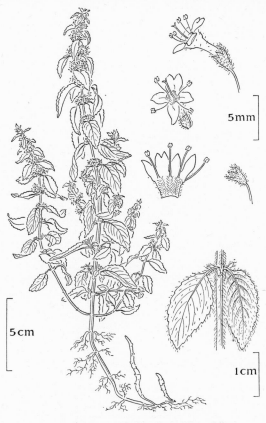

1170. *Mentha arvensis* L. Corn Mint Lilac

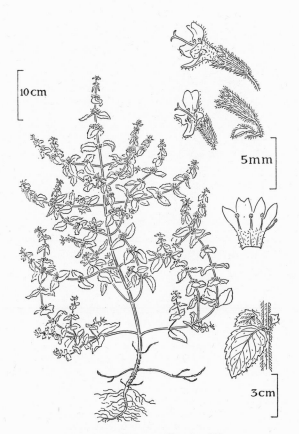

1171. *Mentha × verticillata* L. Lilac

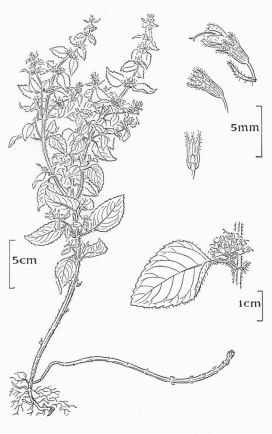

1172. *Mentha × gentilis* L. Lilac

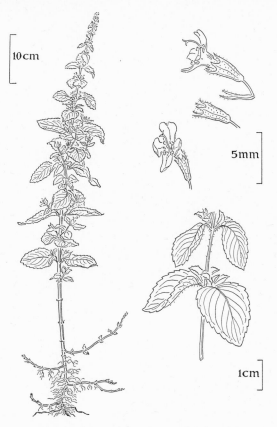

1173. *Mentha* × *smithiana* R. A. Graham   Lilac

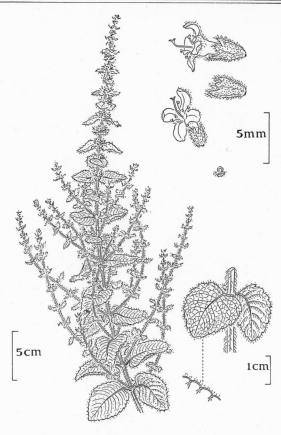

1174. *Mentha* × *muelleriana* F. W. Schultz   Lilac

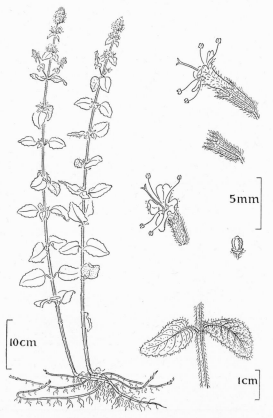

1175. *Mentha aquatica* L.   Water Mint   Lilac

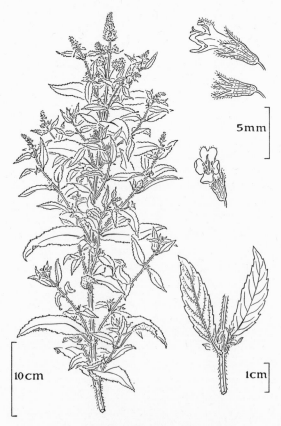

1176. *Mentha* × *piperita* L.   Peppermint
Reddish-lilac

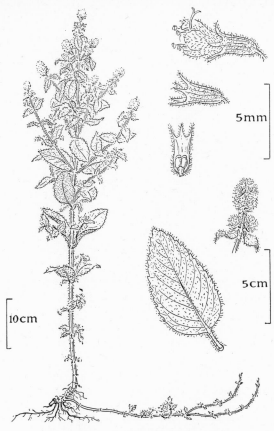

1177. *Mentha × dumetorum* Schultes   Lilac

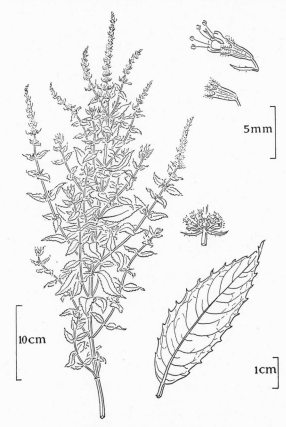

1178. *Mentha spicata* L.   Spearmint   Lilac

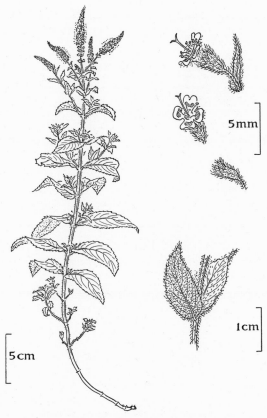

1179. *Mentha longifolia* (L.) Huds.   Horse-mint   Lilac

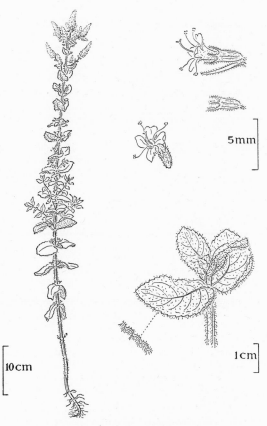

1180. *Mentha × niliaca* Jacq.   Lilac or pinkish-lilac

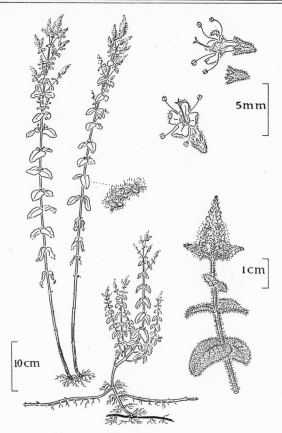

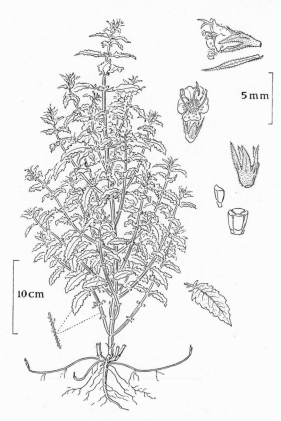

1181. *Mentha rotundifolia* (L.) Huds. Apple-scented mint
Pinkish-lilac

1182. *Lycopus europaeus* L. Gipsy-wort
White and purple

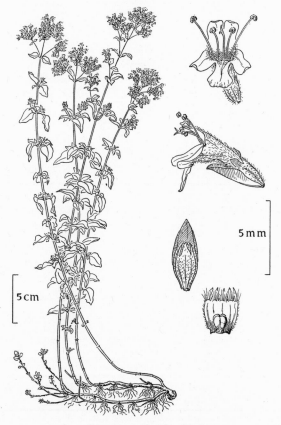

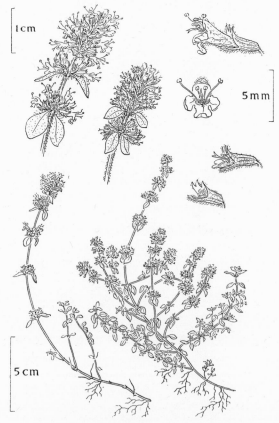

1183. *Origanum vulgare* L. Marjoram
Pinkish-purple

1184. *Thymus pulegioides* L. Larger Wild Thyme
Pinkish-purple

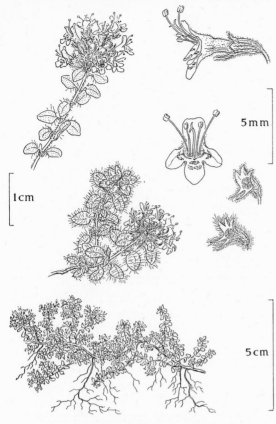

1185.   *Thymus drucei* Ronn.   Wild Thyme
Pinkish-purple

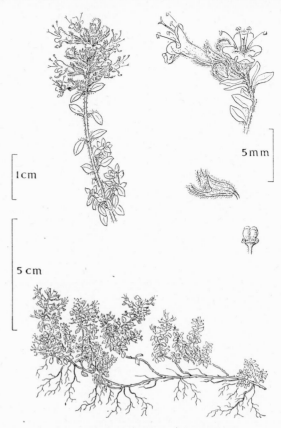

1186.   *Thymus serpyllum* L.   Wild Thyme
Pinkish-purple

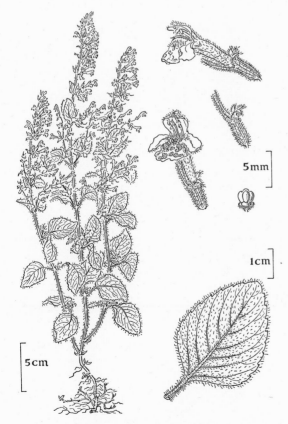

1187.   *Calamintha sylvatica* Bromf.   Wood Calamint
Purplish-pink

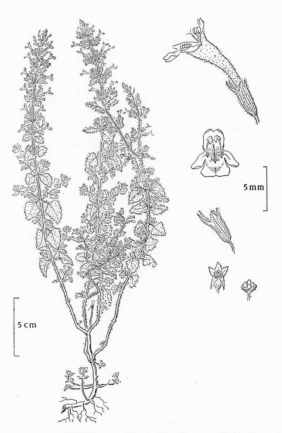

1188.   *Calamintha ascendens* Jord.   Common Calamint
Lilac or purplish

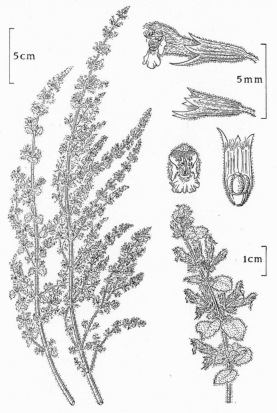

1189.   *Calamintha nepeta* (L.) Savi   Lesser Calamint
Lilac

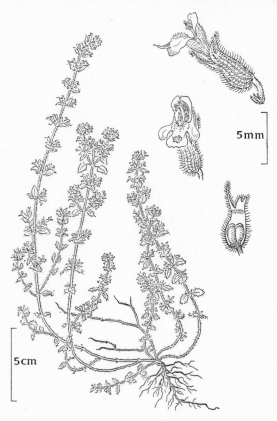

1190.   *Acinos arvensis* (Lam.) Dandy   Basil-thyme
Violet and white

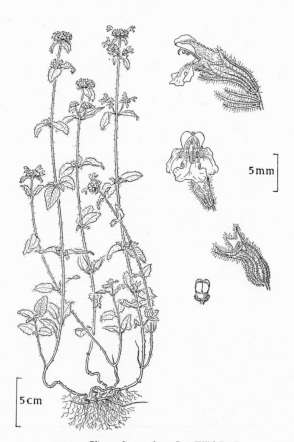

1191.   *Clinopodium vulgare* L.   Wild Basil
Pinkish-purple

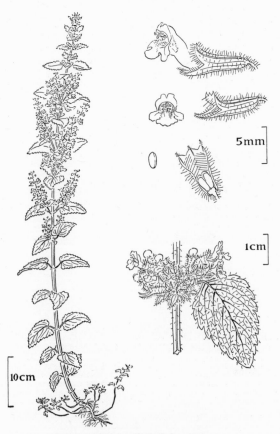

1192.   *Melissa officinalis* L.   Balm
White or pinkish

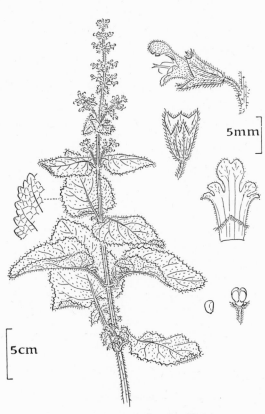

1193. *Salvia verticillata* L.   Violet

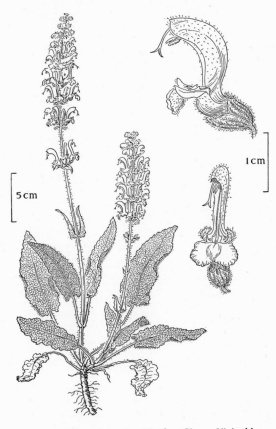

1194. *Salvia pratensis* L.   Meadow Clary   Violet-blue

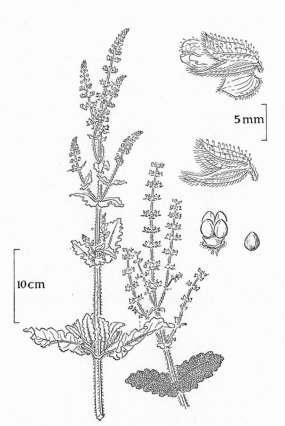

1195. *Salvia horminoides* Pourr.   Wild Clary   Violet-blue

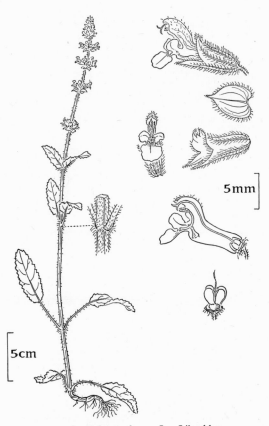

1196. *Salvia verbenaca* L.   Lilac-blue

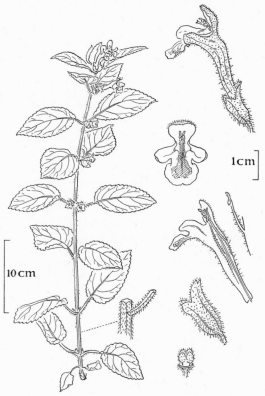

1197. *Melittis melisophyllum* L. Bastard Balm Pink

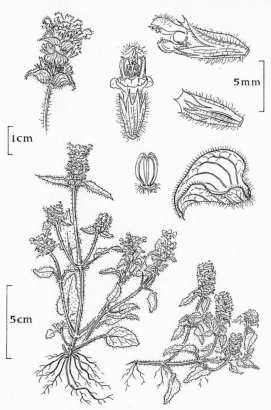

1198. *Prunella vulgaris* L. Self-heal Violet

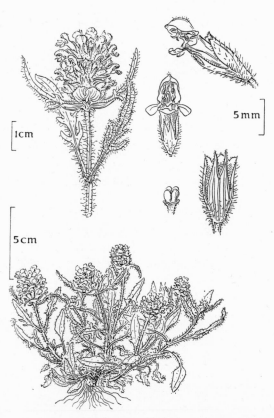

1199. *Prunella laciniata* (L.) L. Cream-white

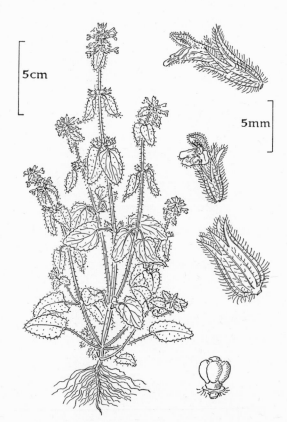

1200. *Stachys arvensis* (L.) L. Field Woundwort
Pale purple

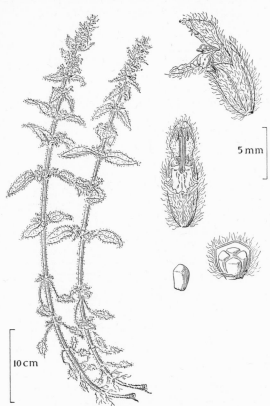

1201.   *Stachys germanica* L.   Downy Woundwort
Pale purplish

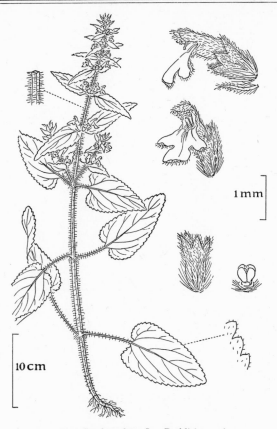

1202.   *Stachys alpina* L.   Reddish-purple

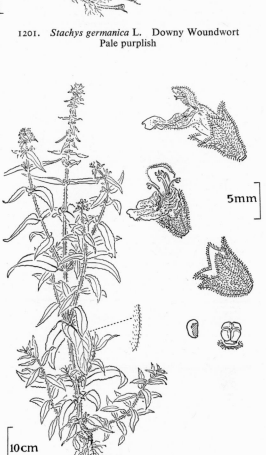

1203.   *Stachys palustris* L.   Marsh Woundwort
Purple

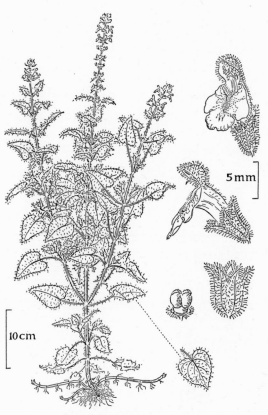

1204.   *Stachys sylvatica* L.   Hedge Woundwort
Purplish

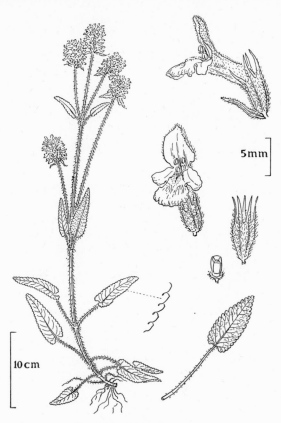

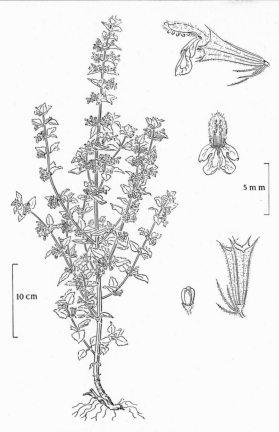

1205. *Betonica officinalis* L. Betony
Reddish-purple

1206. *Ballota nigra* L. Black Horehound Purple

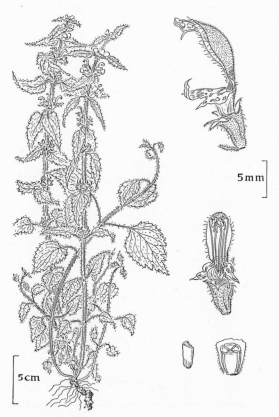

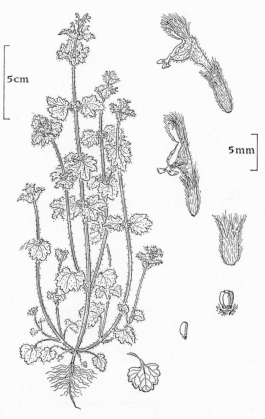

1207. *Galeobdolon luteum* Huds. Yellow Archangel
Yellow

1208. *Lamium amplexicaule* L. Henbit
Pinkish-purple

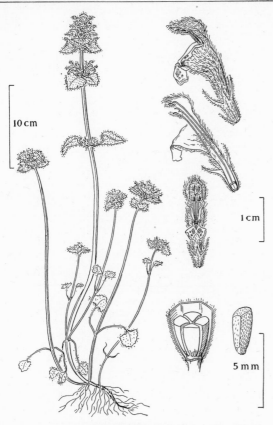

1209.   *Lamium moluccellifolium* Fr.   Intermediate Dead-
             nettle   Pinkish-purple

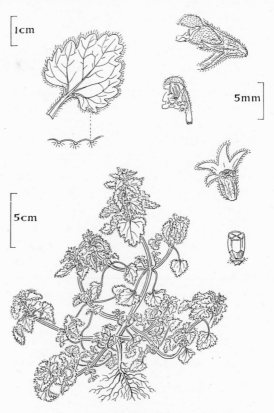

1210.   *Lamium hybridum* Vill.   Cut-leaved Dead-nettle
                        Pinkish-purple

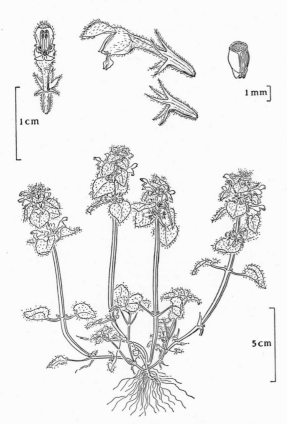

1211.   *Lamium purpureum* L.   Red Dead-nettle
                        Pinkish-purple

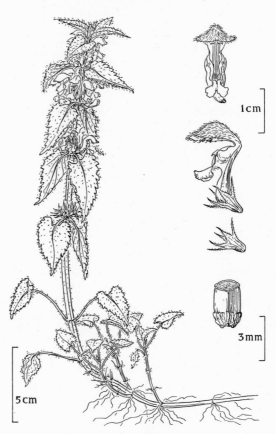

1212.   *Lamium album* L.   White Dead-nettle   White

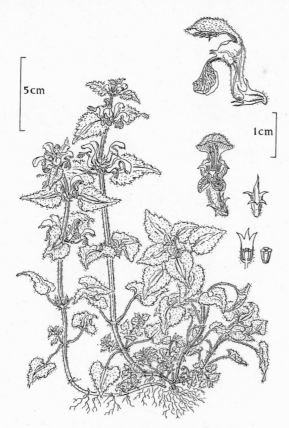

1213. *Lamium maculatum* L.   Spotted Dead-nettle
Pinkish-purple

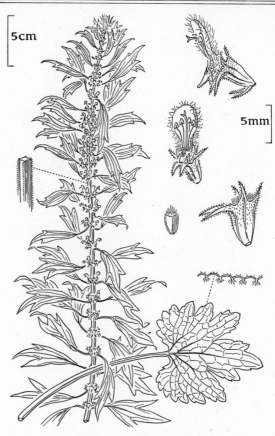

1214. *Leonurus cardiaca* L.   Motherwort
White or pale pink

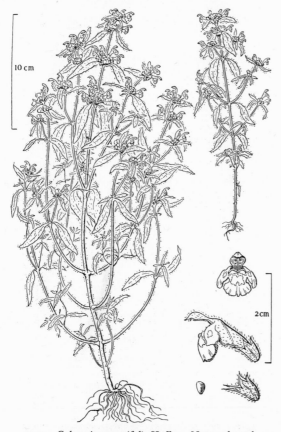

1215. *Galeopsis angustifolia* Hoffm.   Narrow-leaved
Hemp-nettle   Pinkish-purple

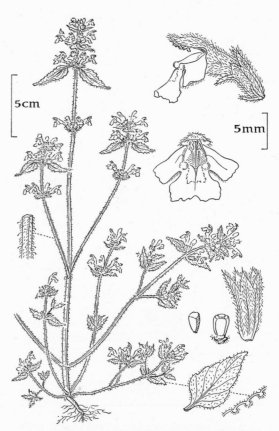

1216. *Galeopsis segetum* Neck.   Downy Hemp-nettle
Pale yellow

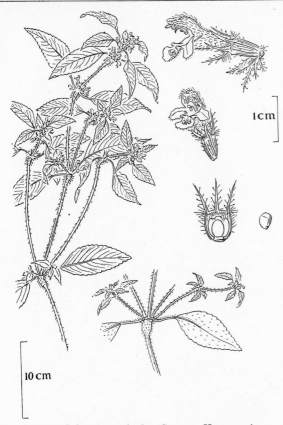

1217. *Galeopsis tetrahit* L.   Common Hemp-nettle
Purple, pink or white

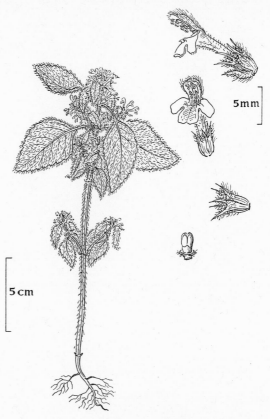

1218. *Galeopsis bifida* Boenn.   Common Hemp-nettle
Purple, pink or white

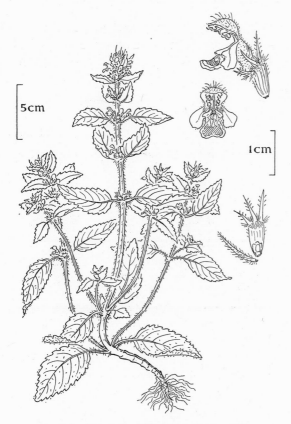

1219. *Galeopsis speciosa* Mill.   Large-flowered Hemp-
nettle   Pale yellow and violet

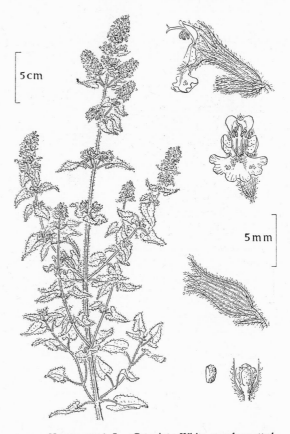

1220. *Nepeta cataria* L.   Cat-mint   White, purple-spotted

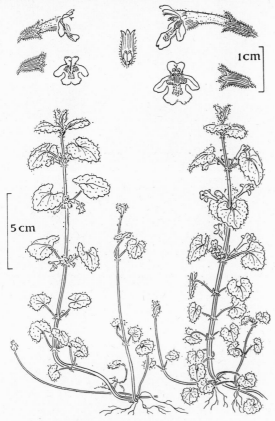

1221.   *Glechoma hederacea* L.   Ground Ivy   Violet

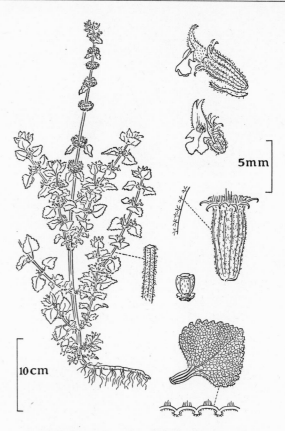

1222.   *Marrubium vulgare* L.   White Horehound   Whitish

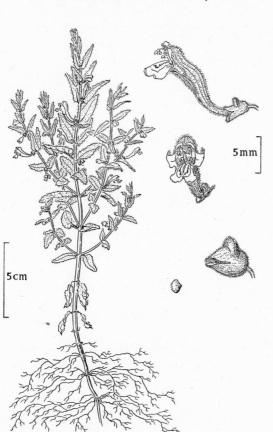

1223.   *Scutellaria galericulata* L.   Skull-cap   Blue-violet

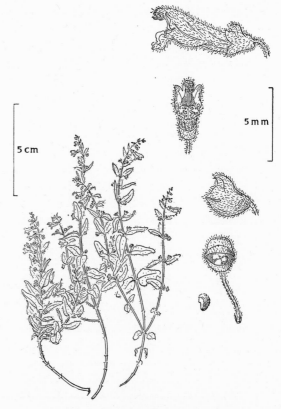

1224.   *Scutellaria minor* Huds.   Lesser Skull-cap
Pinkish-purple

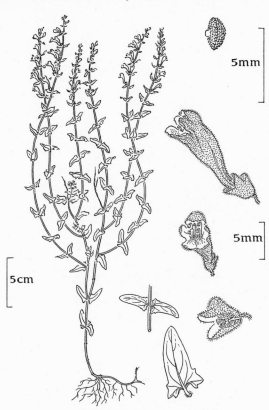

5mm

5mm

5cm

1225. *Scutellaria hastifolia* L.   Blue-violet

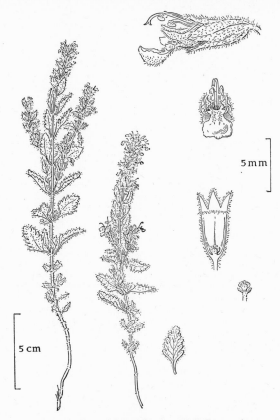

5 mm

5 cm

1226. *Teucrium chamaedrys* L.   Wall Germander
Pinkish-purple

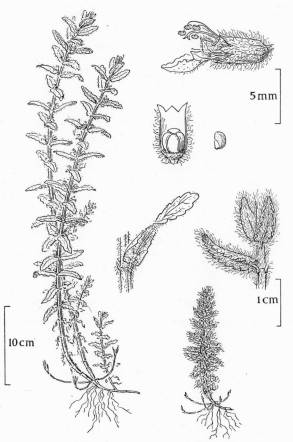

5 mm

1 cm

10 cm

1227. *Teucrium scordium* L.   Water Germander   Purple

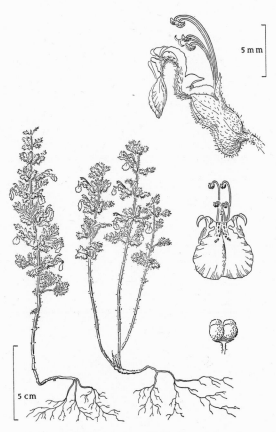

5 mm

5 cm

1228. *Teucrium botrys* L.   Cut-leaved Germander
Pinkish-purple

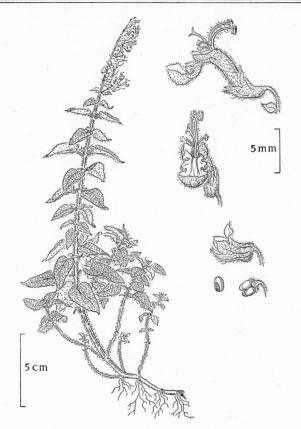

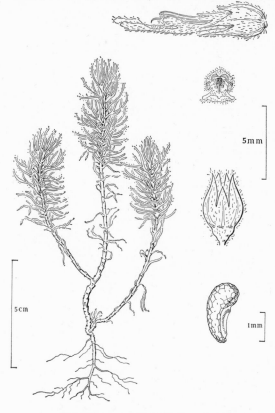

1229.   *Teucrium scorodonia* L.   Wood Sage
Yellowish-green

1230.   *Ajuga chamaepitys* (L.) Schreb.   Ground-pine
Yellow

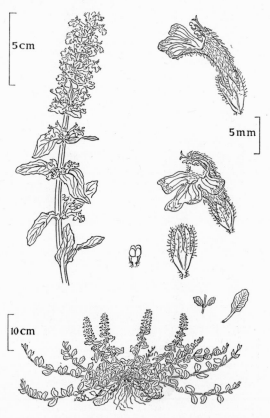

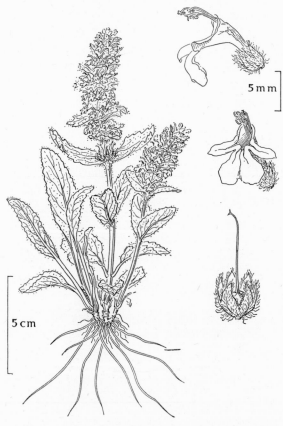

1231.   *Ajuga reptans* L.   Bugle   Blue

1232.   *Ajuga genevensis* L.   Blue

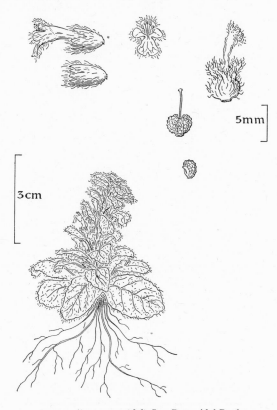

1233. *Ajuga pyramidalis* L. Pyramidal Bugle
Pale violet-blue

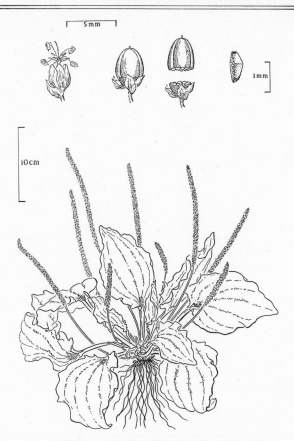

1234. *Plantago major* L. Great Plantain Yellowish

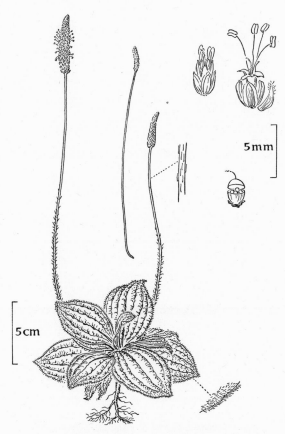

1235. *Plantago media* L. Hoary Plantain Whitish

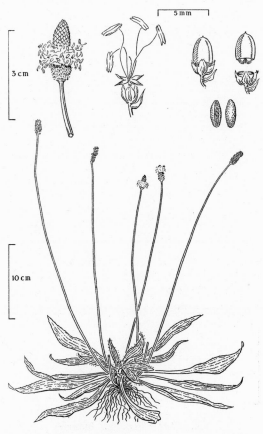

1236. *Plantago lanceolata* L. Ribwort Brownish

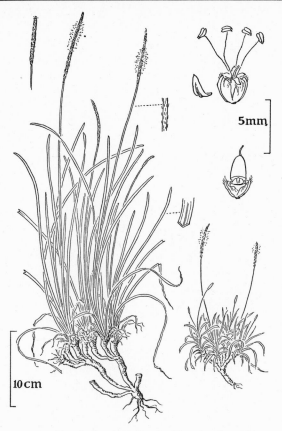

1237.   *Plantago maritima* L.   Sea Plantain   Brownish

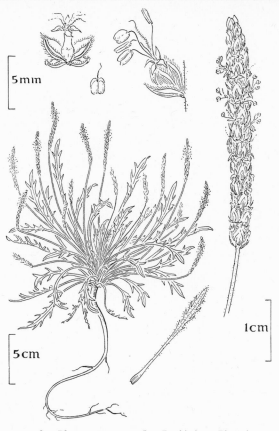

1238.   *Plantago coronopus* L.   Buck's-horn Plantain
Brownish

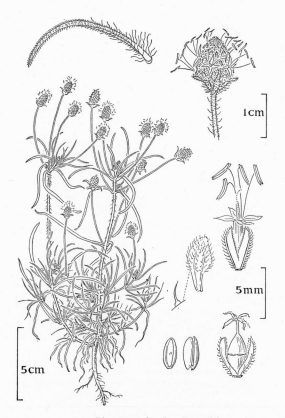

1239.   *Plantago indica* L.   Brownish

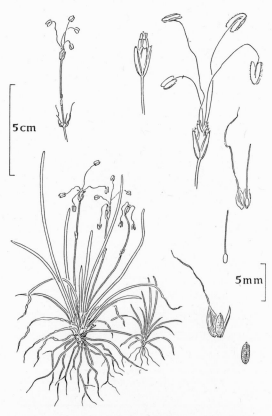

1240.   *Littorella uniflora* (L.) Aschers.   Shore-weed
Brownish

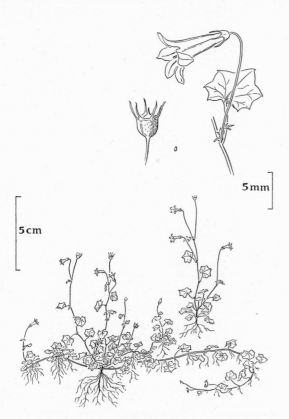

1241. *Wahlenbergia hederacea* (L.) Rchb.   Ivy Campanula
Pale blue

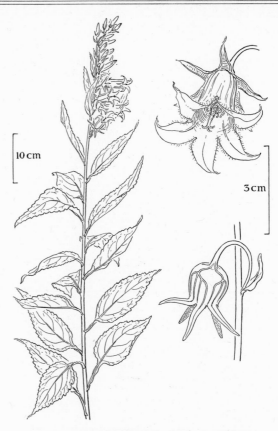

1242. *Campanula latifolia* L.   Large Campanula
Blue-purple

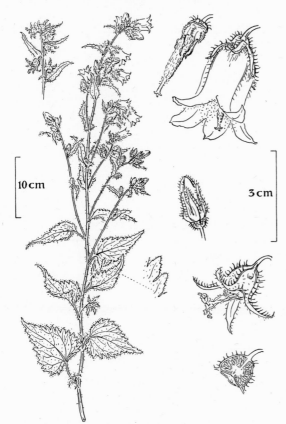

1243. *Campanula trachelium* L.   Bats-in-the-Belfry
Blue-purple

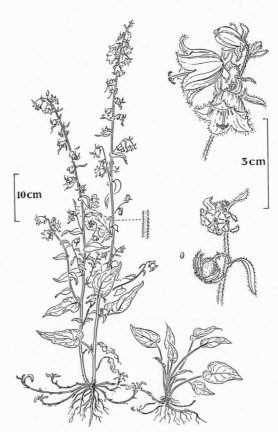

1244. *Campanula rapunculoides* L.   Creeping Campanula
Blue-purple

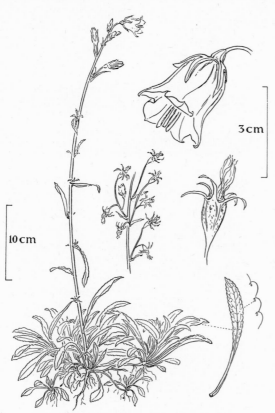

1245.   *Campanula persicifolia* L.   Blue or white

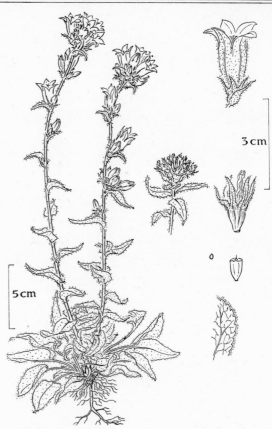

1246.   *Campanula glomerata* L.   Clustered Bellflower
Blue-purple

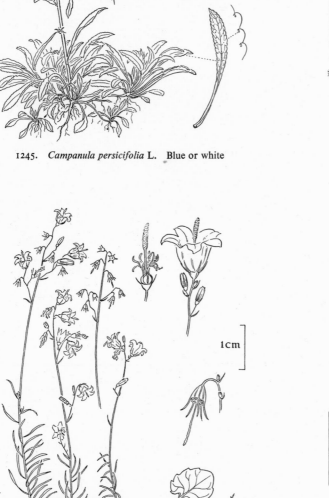

1247.   *Campanula rotundifolia* L.   Harebell   Blue

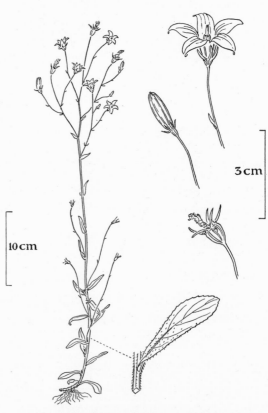

1248.   *Campanula patula* L.   Purple

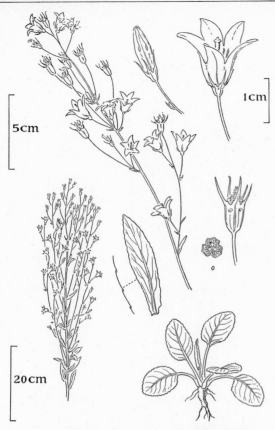

1249.   *Campanula rapunculus* L.   Rampion   Purple

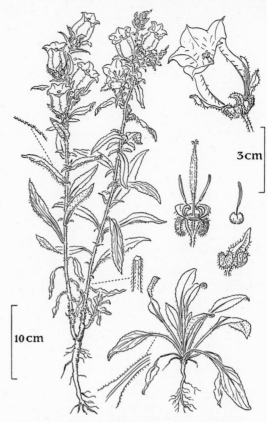

1250.   *Campanula medium* L.   Canterbury Bell
Violet-blue

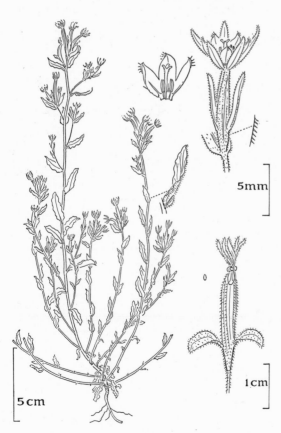

1251.   *Legousia hybrida* (L.) Delarb.   Venus's Looking-
glass   Reddish-purple or lilac

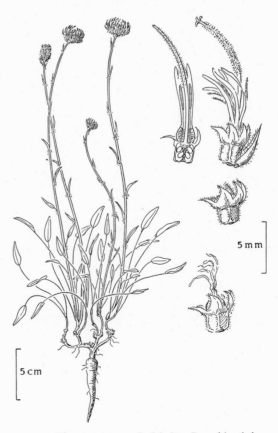

1252.   *Phyteuma tenerum* R. Schultz   Round-headed
Rampion   Violet

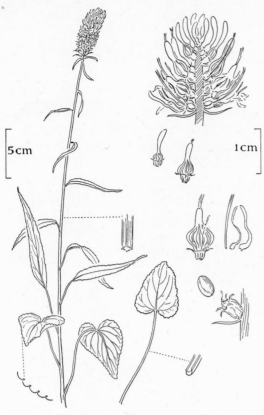

1253. *Phyteuma spicatum* L. Yellowish

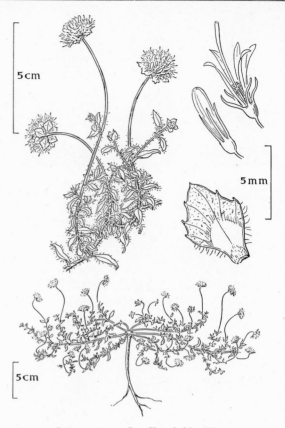

1254. *Jasione montana* L. Sheep's-bit Blue

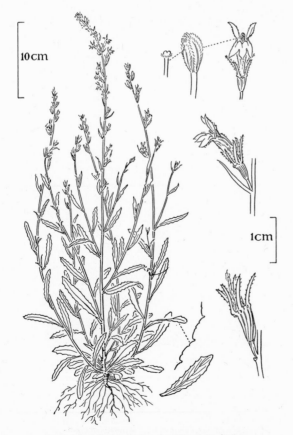

1255. *Lobelia urens* L. Acrid Lobelia Blue

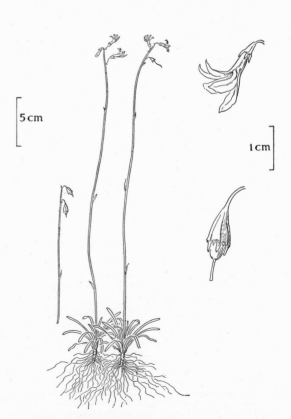

1256. *Lobelia dortmanna* L. Water Lobelia Pale lilac

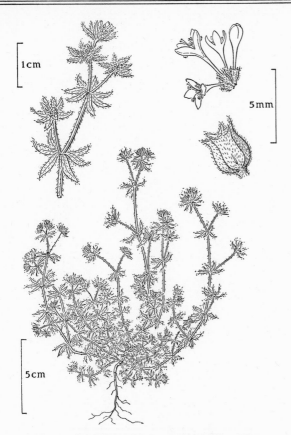

1257. *Sherardia arvensis* L.   Field Madder   Lilac

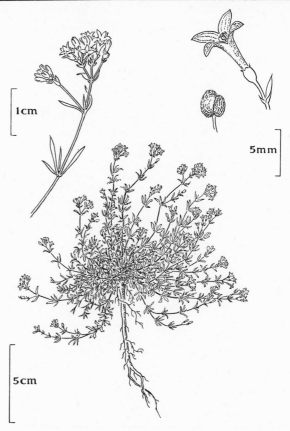

1258. *Asperula cynanchica* L.   Squinancy Wort   Whitish

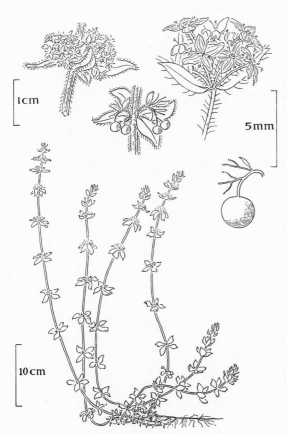

1259. *Galium cruciata* (L.) Scop.   Crosswort, Mugwort
Yellow

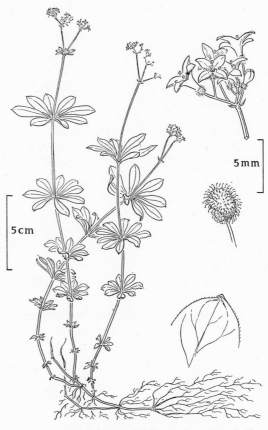

1260. *Galium odoratum* (L.) Scop.   Sweet Woodruff
White

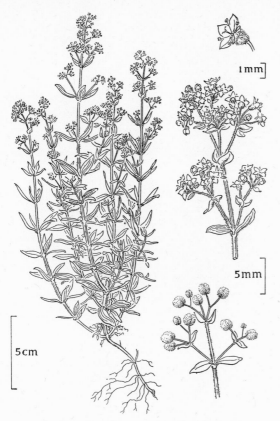

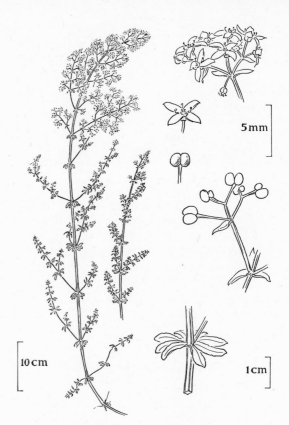

1261. *Galium boreale* L. Northern Bedstraw White

1262. *Galium mollugo* L. Hedge Bedstraw White

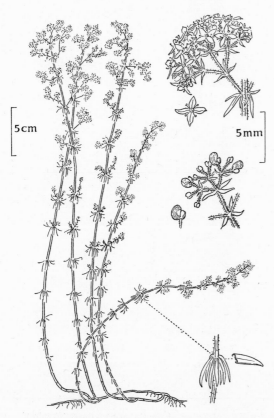

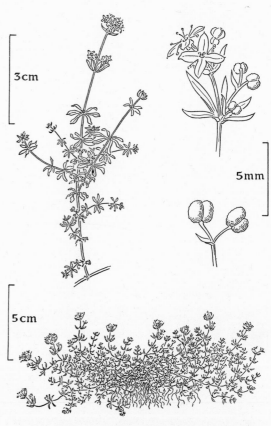

1263. *Galium verum* L. Lady's Bedstraw Yellow

1264. *Galium saxatile* L. Heath Bedstraw White

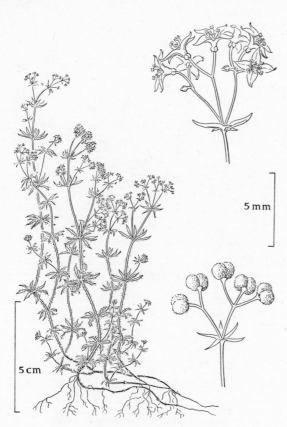

1265.   *Galium sterneri* Ehrendorf.   Cream

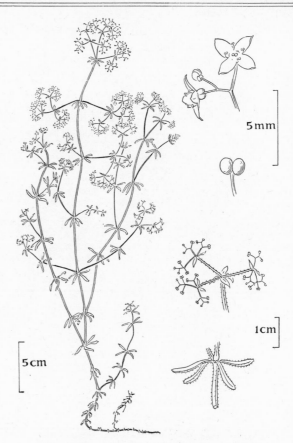

1266.   *Galium palustre* L.   Marsh Bedstraw   White

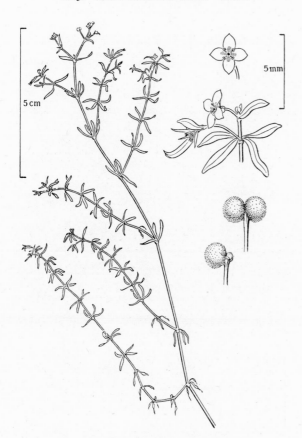

1267.   *Galium debile* Desv.   Slender Bedstraw
        Pinkish-white

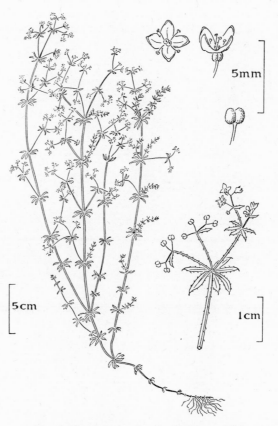

1268.   *Galium uliginosum* L.   Fen Bedstraw   White

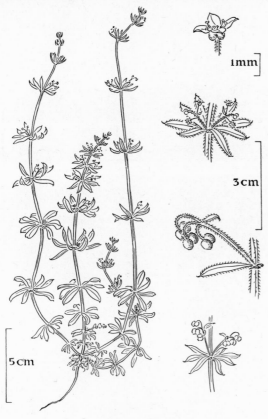

1269. *Galium tricornutum* Dandy   Rough Corn
Bedstraw   Cream

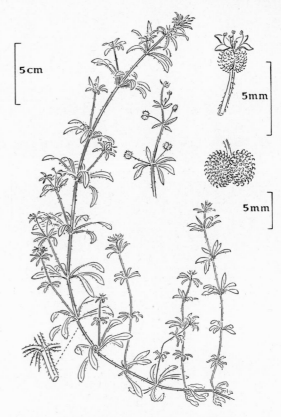

1270. *Galium aparine* L.   Goosegrass   Whitish

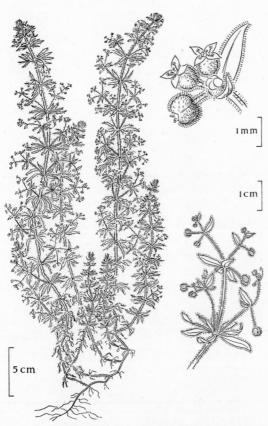

1271. *Galium spurium* L.   False Cleavers   Whitish

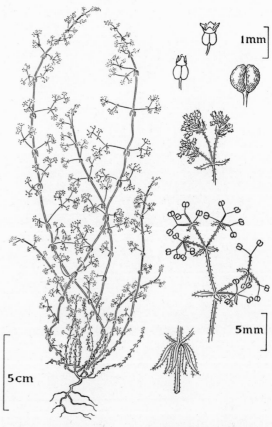

1272. *Galium parisiense* L.   Wall Bedstraw   Greenish

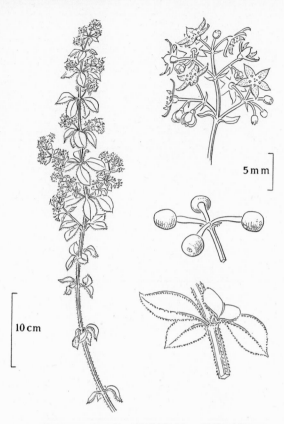

1273. *Rubia peregrina* L.   Wild Madder   Yellowish-green

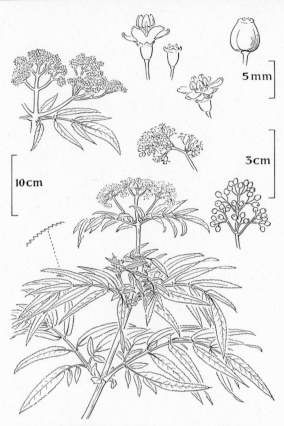

1274. *Sambucus ebulus* L.   Danewort   White

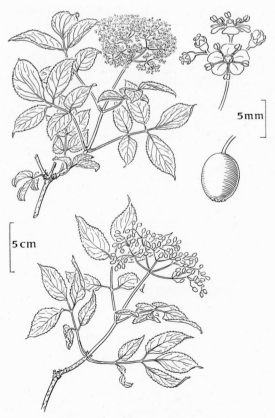

1275. *Sambucus nigra* L.   Elder   Cream

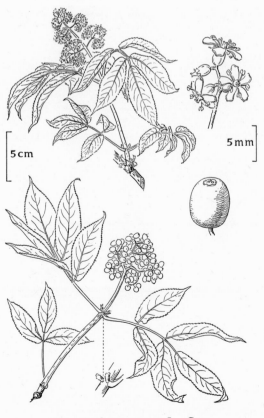

1276. *Sambucus racemosa* L.   Cream

5·2

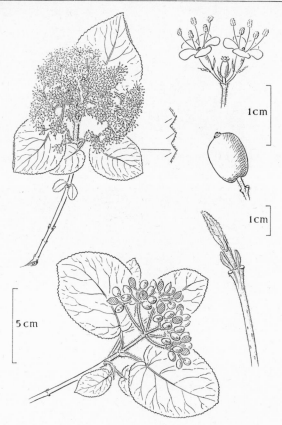

1277.   *Viburnum lantana* L.   Wayfaring Tree   Cream

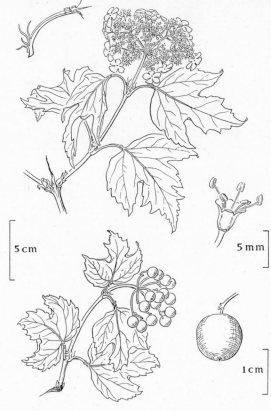

1278.   *Viburnum opulus* L.   Guelder Rose   White

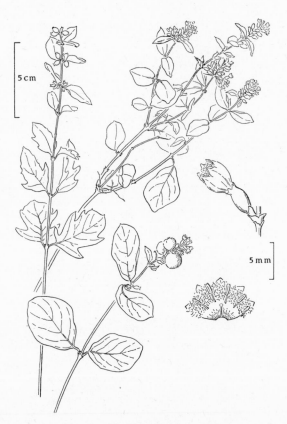

1279.   *Symphoricarpos rivularis* Suksdorf
Snowberry   Pink

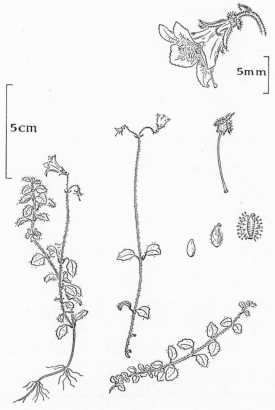

1280.   *Linnaea borealis* L.   Linnaea   Pink

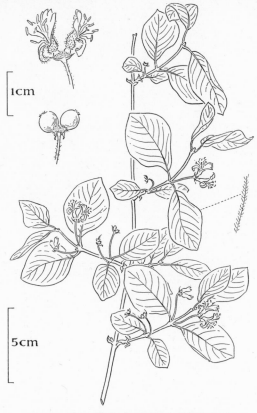

1281. *Lonicera xylosteum* L.   Fly Honeysuckle
Yellowish

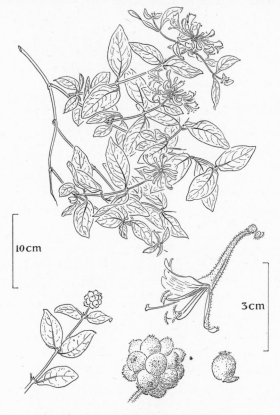

1282. *Lonicera periclymenum* L.   Honeysuckle
Cream, purplish or yellowish

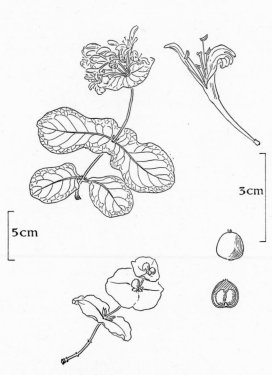

1283. *Lonicera caprifolium* L.   Perfoliate Honeysuckle
Cream, purplish or yellowish

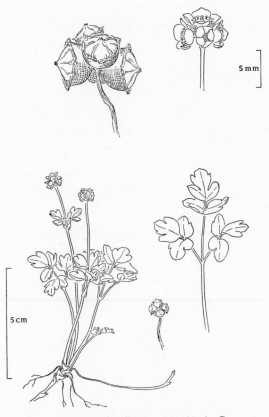

1284. *Adoxa moschatellina* L.   Moschatel   Green

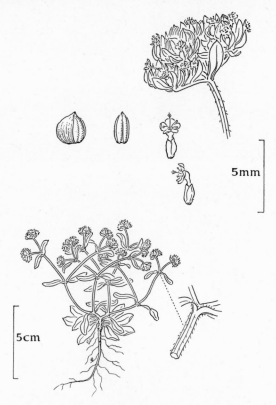

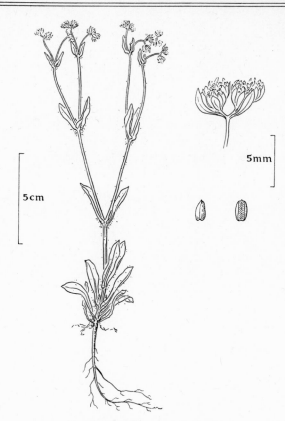

1285.   *Valerianella locusta* (L.) Betcke   Lamb's Lettuce,
Corn Salad   Lilac

1286.   *Valerianella carinata* Lois.   Lilac

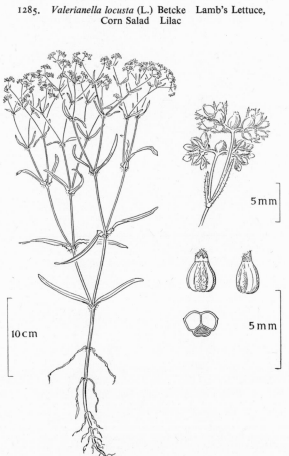

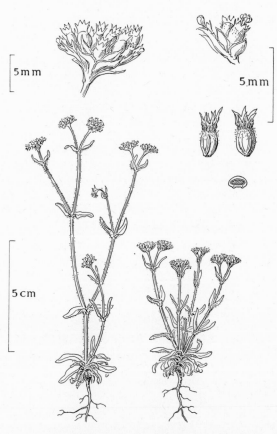

1287.   *Valerianella rimosa* Bast.   Lilac

1288.   *Valerianella eriocarpa* Desv.   Lilac

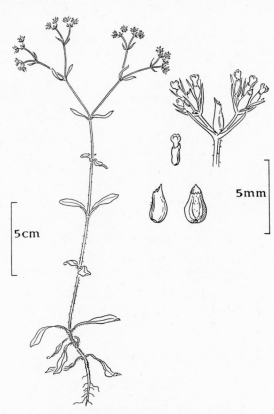

1289.   *Valerianella dentata* (L.) Poll.   Lilac

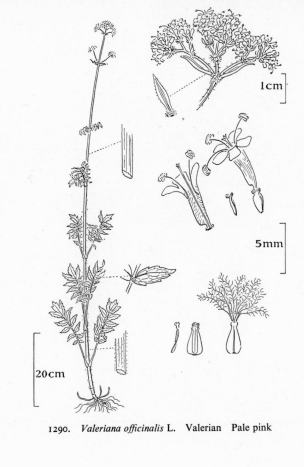

1290.   *Valeriana officinalis* L.   Valerian   Pale pink

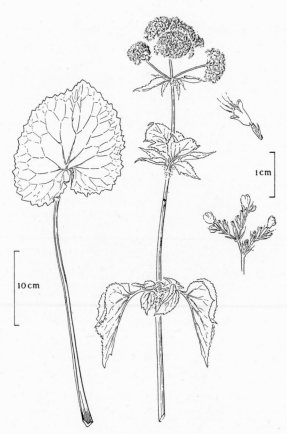

1291.   *Valeriana pyrenaica* L.   Pyrenean Valerian   Pale pink

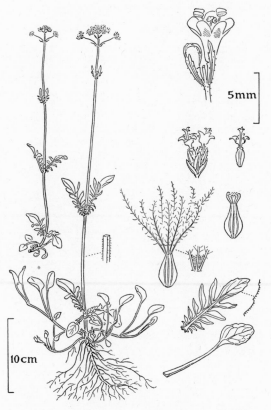

1292.   *Valeriana dioica* L.   Marsh Valerian   Pinkish

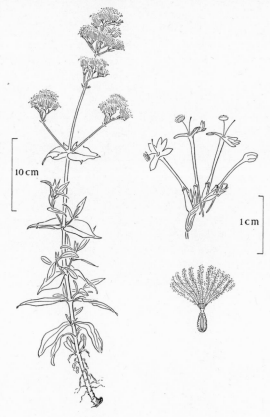

1293.　*Centranthus ruber* (L.) DC.　Red Valerian
Red or white

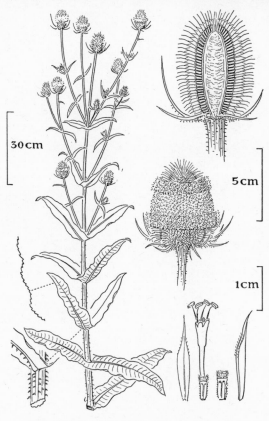

1294.　*Dipsacus fullonum* L.　Teasel　Pink-purple

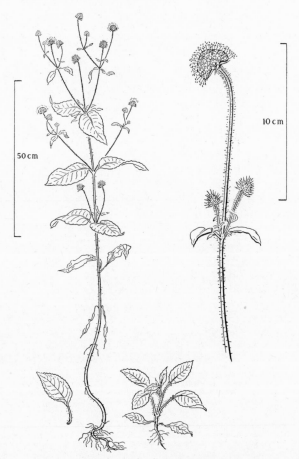

1295.　*Dipsacus pilosus* L.　Small Teasel　Whitish

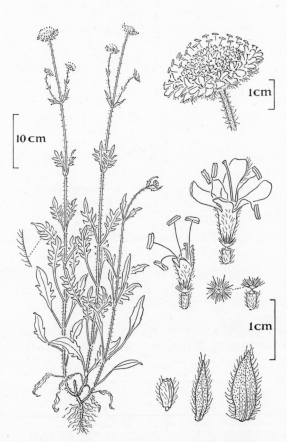

1296.　*Knautia arvensis* (L.) Coult.　Field Scabious
Bluish-lilac

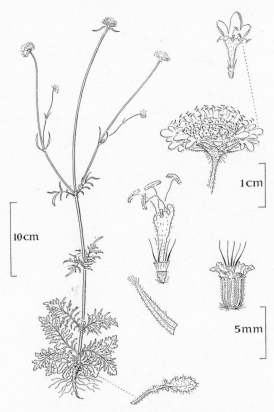

1297. *Scabiosa columbaria* L.   Small Scabious
Bluish-lilac

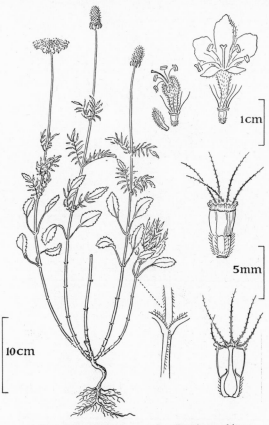

1298. *Scabiosa atropurpurea* L.   Purple to white

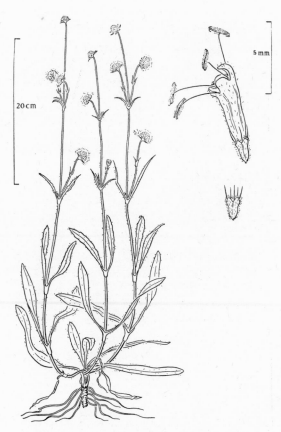

1299. *Succisa pratensis* Moench   Devil's-bit Scabious
Purple

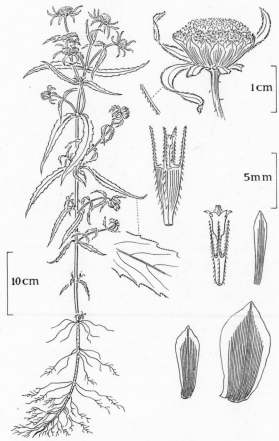

1300. *Bidens cernua* L.   Nodding Bur-Marigold   Yellow

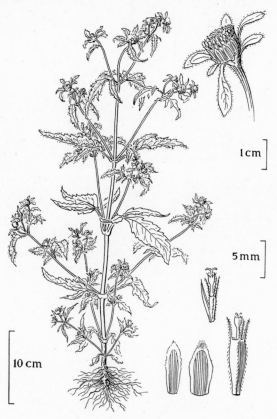

1301. *Bidens tripartita* L. Tripartite Bur-Marigold
Yellow

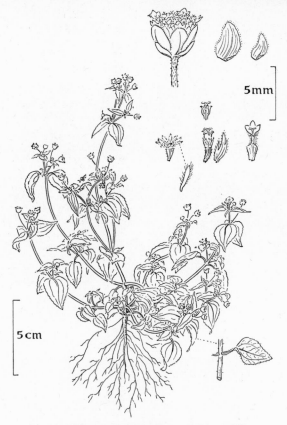

1302. *Galinsoga parviflora* Cav. Gallant Soldier
White and yellow

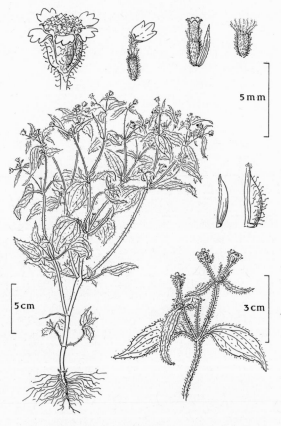

1303. *Galinsoga ciliata* (Raf.) Blake White and yellow

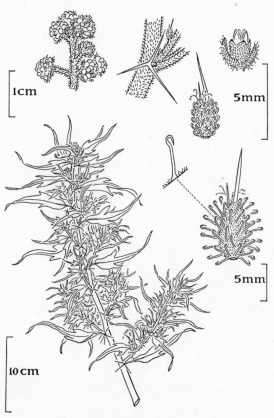

1304. *Xanthium spinosum* L. Spiny Cocklebur

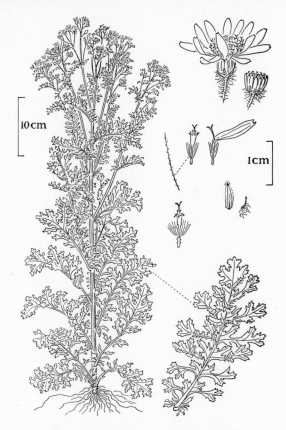

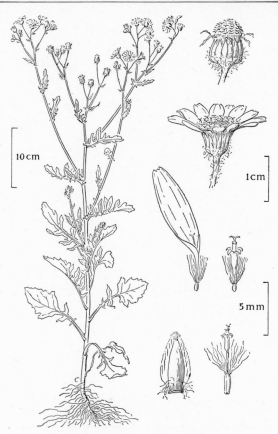

1305.   *Senecio jacobaea* L.   Ragwort   Yellow

1306.   *Senecio aquaticus* Hill   Marsh Ragwort   Yellow

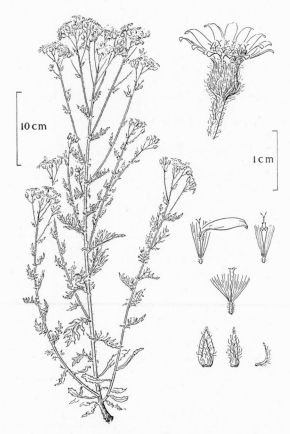

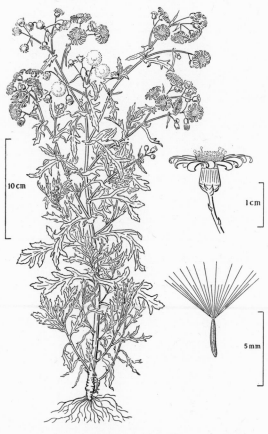

1307.   *Senecio erucifolius* L.   Hoary Ragwort   Yellow

1308.   *Senecio squalidus* L.   Oxford Ragwort   Yellow

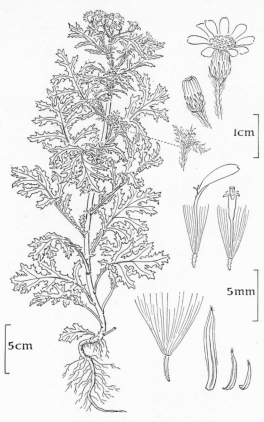

1309.   *Senecio cambrensis* Rosser   Yellow

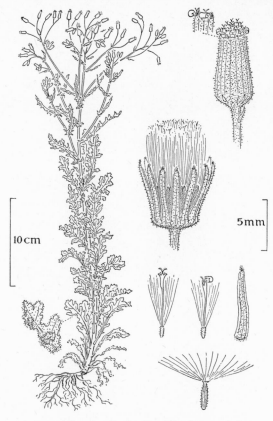

1310.   *Senecio sylvaticus* L.   Wood Groundsel
Yellow

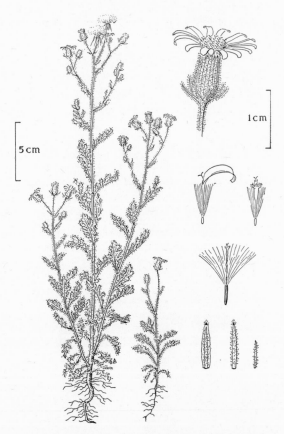

1311.   *Senecio viscosus* L.   Stinking Groundsel   Yellow

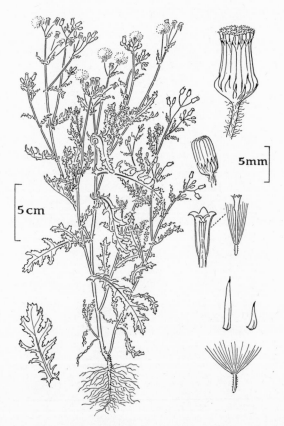

1312.   *Senecio vulgaris* L.   Groundsel   Yellow

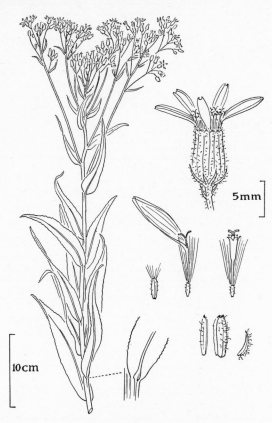

1313.   *Senecio doria* L.   Yellow

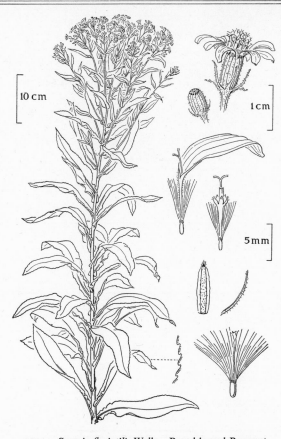

1314.   *Senecio fluviatilis* Wallr.   Broad-leaved Ragwort
Yellow

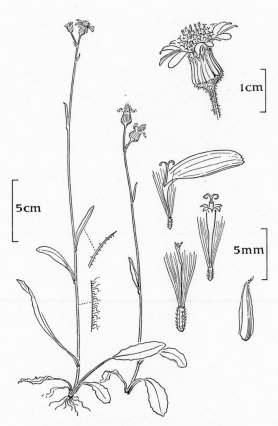

1315.   *Senecio integrifolius* (L.) Clairv.   Yellow

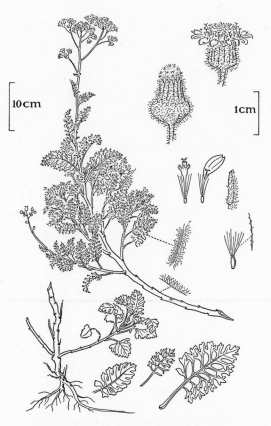

1316.   *Senecio cineraria* DC.   Cineraria   Yellow

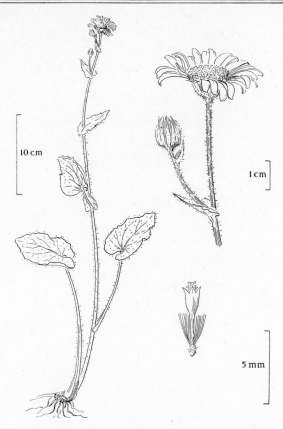

1317.   *Doronicum pardalianches* L.   Great Leopard's-bane
        Yellow

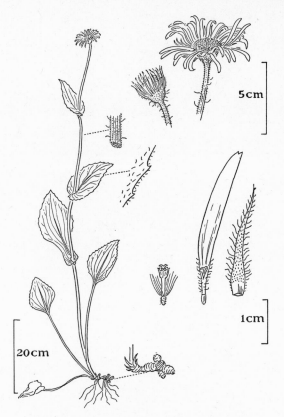

1318.   *Doronicum plantagineum* L.   Leopard's-bane
        Yellow

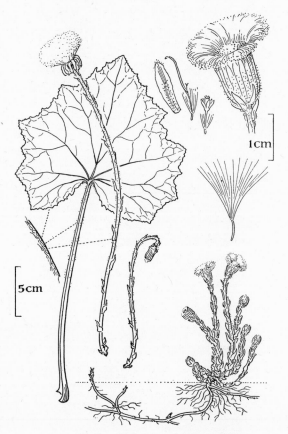

1319.   *Tussilago farfara* L.   Coltsfoot   Yellow

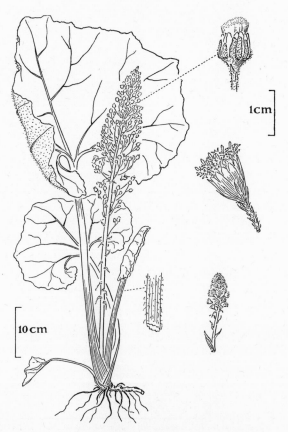

1320.   *Petasites hybridus* (L.) Gaertn., Mey. & Scherb.
        Butterbur   Purplish

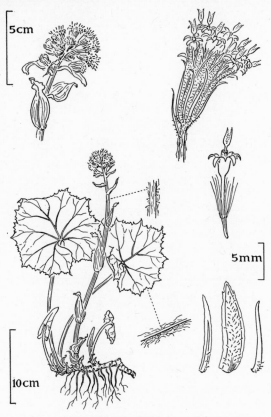

1321. *Petasites albus* (L.) Gaertn. White Butterbur
Whitish

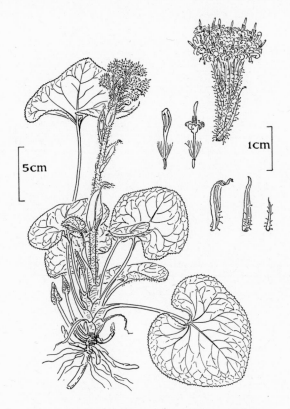

1322. *Petasite fragrans* (Vill.) C. Presl Winter Heliotrope
Lilac

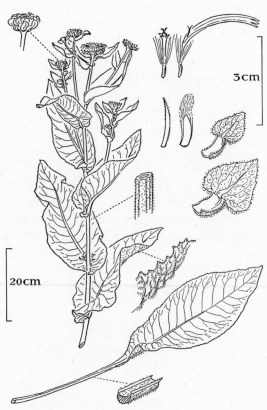

1323. *Inula helenium* L. Elecampane Yellow

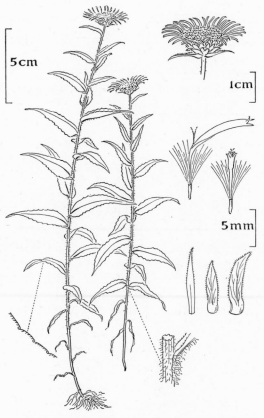

1324. *Inula salicina* L. Willow-leaved Inula Yellow

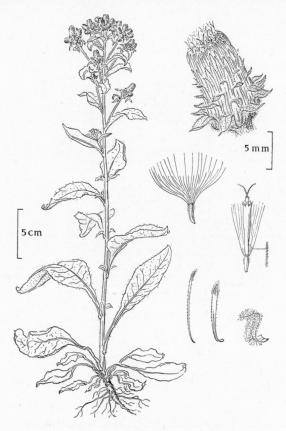

1325. *Inula conyza* DC. Ploughman's Spikenard
Yellowish

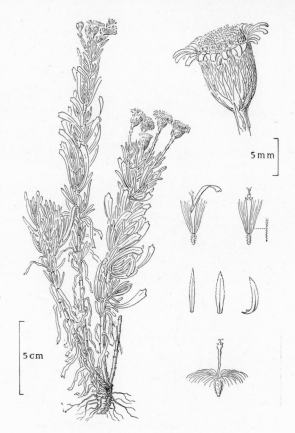

1326. *Inula crithmoides* L. Golden Samphire Yellow

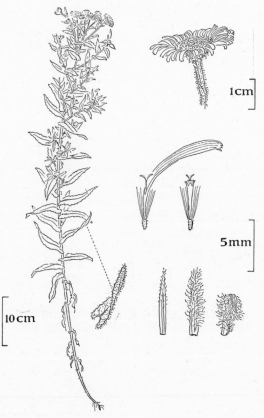

1327. *Pulicaria dysenterica* (L.) Bernh. Fleabane Yellow

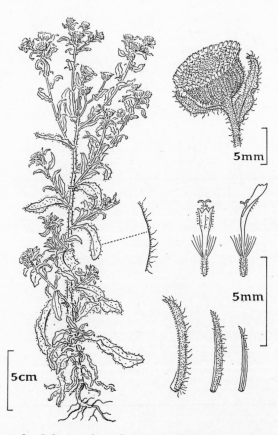

1328. *Pulicaria vulgaris* Gaertn. Small Fleabane Yellow

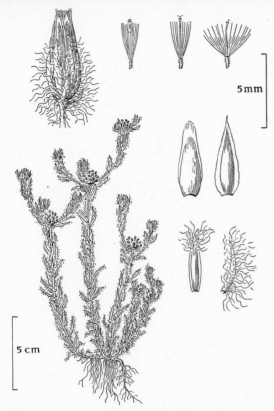

1329.   *Filago germanica* (L.) L.   Cudweed   Yellow

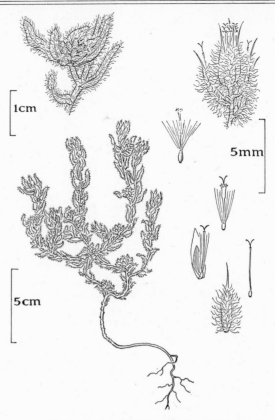

1330.   *Filago apiculata* G. E. Sm.   Red-tipped Cudweed
                      Yellow

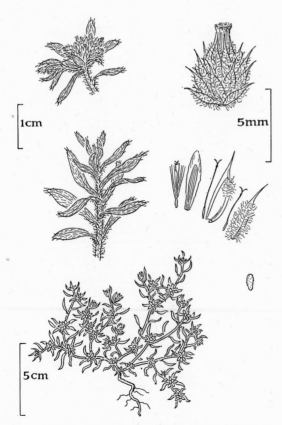

1331.   *Filago spathulata* C. Presl   Spathulate Cudweed
                  Yellowish

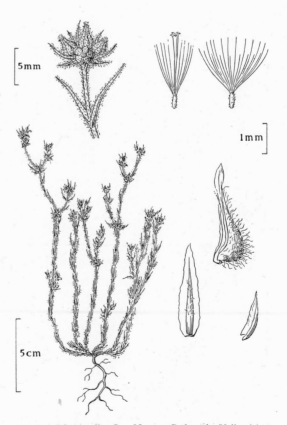

1332.   *Filago gallica* L.   Narrow Cudweed   Yellowish

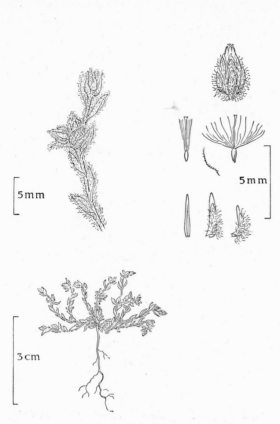

1333. *Filago minima* (Sm.) Pers.   Slender Cudweed
Yellowish

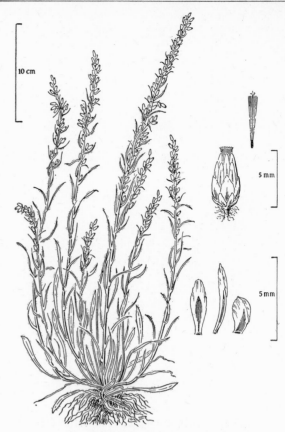

1334. *Gnaphalium sylvaticum* L.   Wood Cudweed
Pale brown

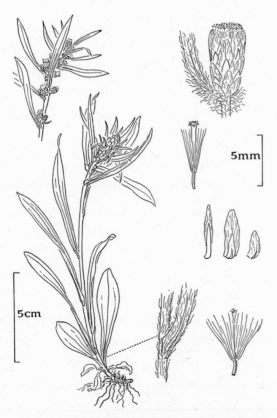

1335. *Gnaphalium norvegicum* Gunn.   Highland Cudweed
Pale brown

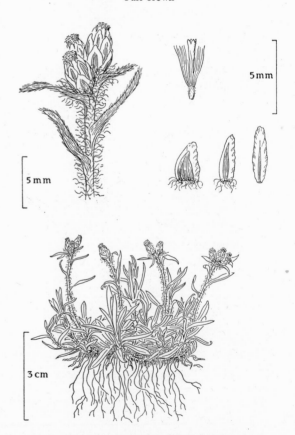

1336. *Gnaphalium supinum* L.   Dwarf Cudweed
Pale brown

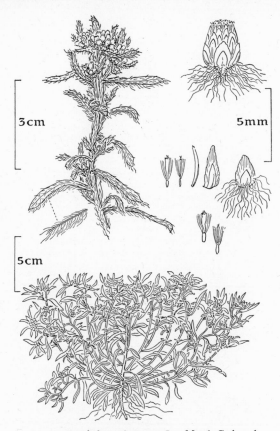

1337.   *Gnaphalium uliginosum* L.   Marsh Cudweed
                 Yellowish

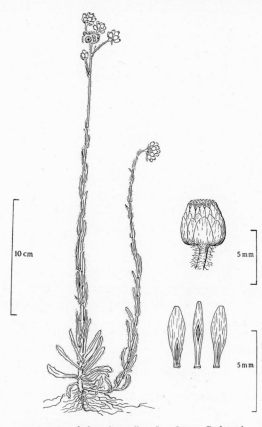

1338.   *Gnaphalium luteoalbum* L.   Jersey Cudweed
                 Yellowish

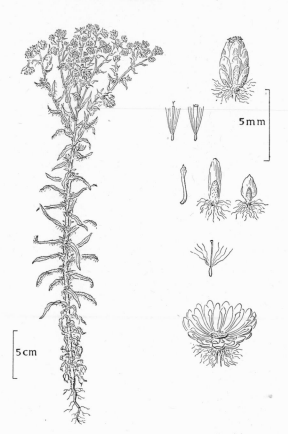

1339.   *Gnaphalium undulatum* L.   Yellowish

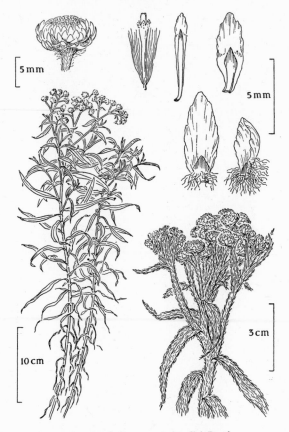

1340.   *Anaphalis margaritacea* (L.) Benth.
              Pearly Everlasting   Yellowish

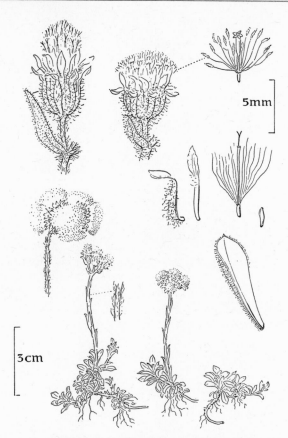

1341. *Antennaria dioica* (L.) Gaertn. Cat's-foot Whitish

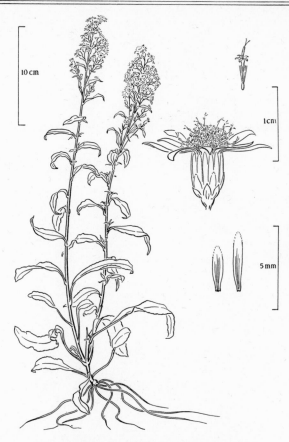

1342. *Solidago virgaurea* L. Golden-rod Yellow

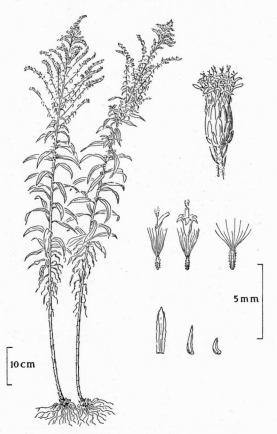

1343. *Solidago canadensis* L. Golden-rod Yellow

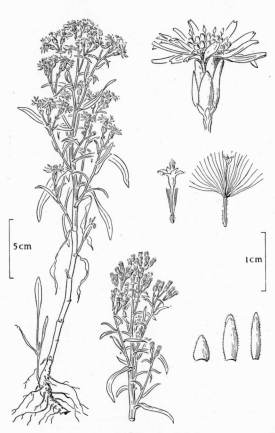

1344. *Aster tripolium* L. Sea Aster
Bluish-purple and yellow

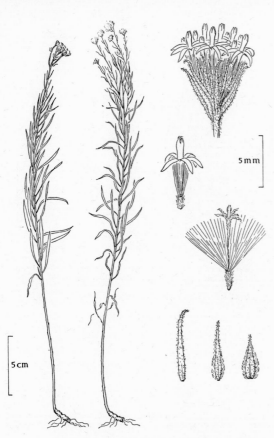

1345.   *Aster linosyris* (L.) Bernh.   Goldilocks   Yellow

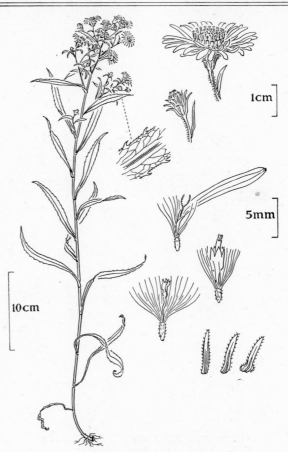

1346.   *Aster salignus* Willd.   Michaelmas Daisy
Bluish-purple and yellow

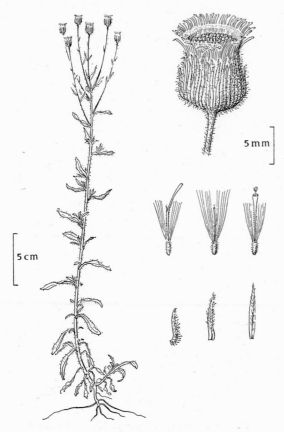

1347.   *Erigeron acer* L.   Blue Fleabane
Pale purple and yellow

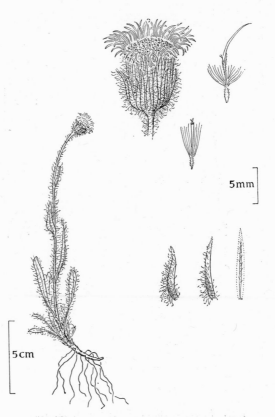

1348.   *Erigeron borealis* (Vierh.) Simmons   Boreal Fleabane
Purple and yellow

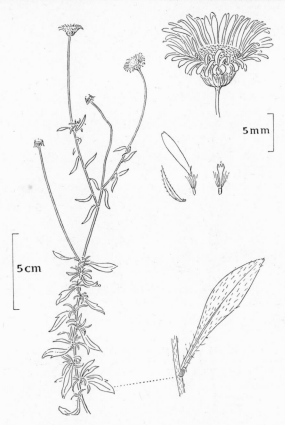

1349. *Erigeron mucronatus* DC. Purple and white

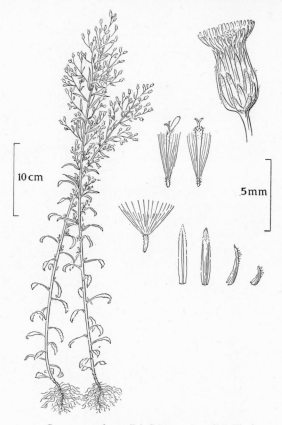

1350. *Conyza canadensis* (L.) Cronq. Canadian Fleabane
Whitish

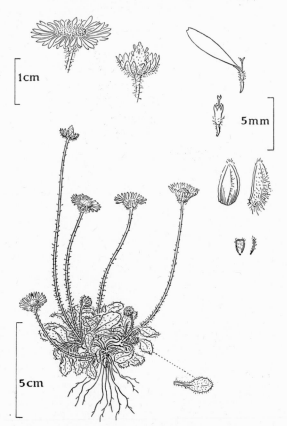

1351. *Bellis perennis* L. Daisy White and yellow

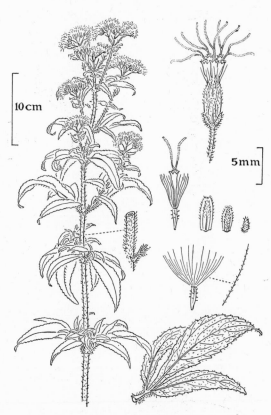

1352. *Eupatorium cannabinum* L. Hemp Agrimony
Purplish or whitish

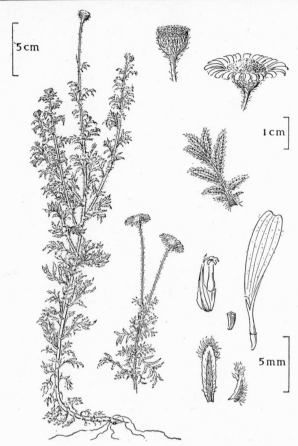

1353.   *Anthemis tinctoria* L.   Yellow Chamomile   Yellow

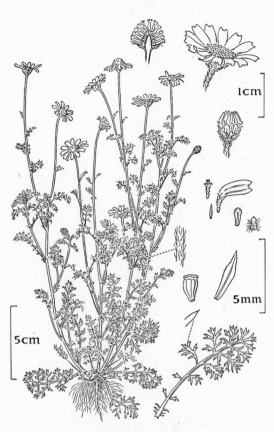

1354.   *Anthemis arvensis* L.   Corn Chamomile
White and yellow

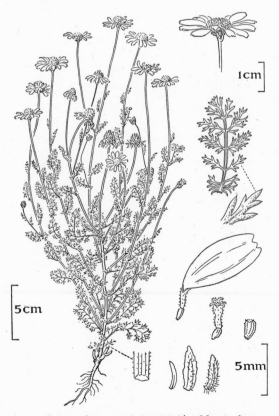

1355.   *Anthemis cotula* L.   Stinking Mayweed
White and yellow

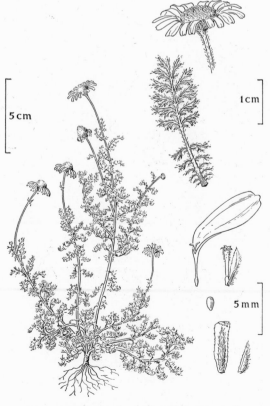

1356.   *Chamaemelum nobile* (L.) All.   Chamomile
White and yellow

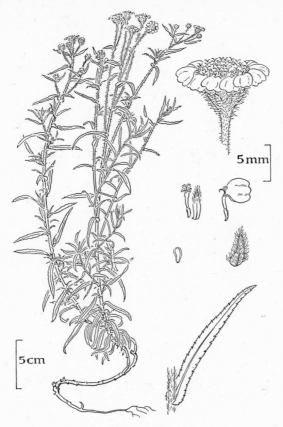

1357. *Achillea ptarmica* L. Sneezewort
White and greenish

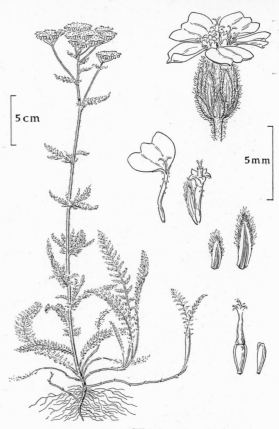

1358. *Achillea millefolium* L. Yarrow White or pink

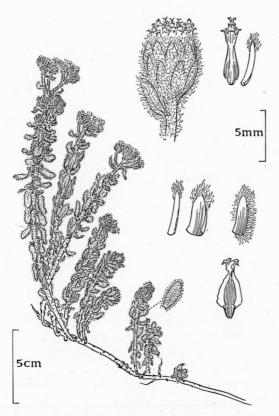

1359. *Otanthus maritimus* (L.) Hoffmans. & Link
Cotton-weed Yellow

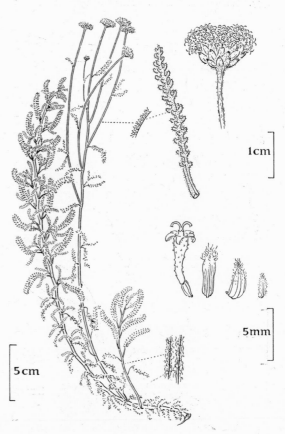

1360. *Santolina chamaecyparissus* L. Lavender Cotton
Yellow

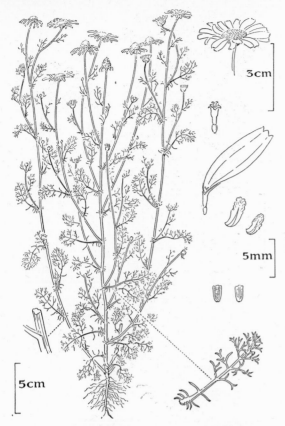

1361.   *Tripleurospermum maritimum* (L.) Koch
Scentless Mayweed   White and yellow

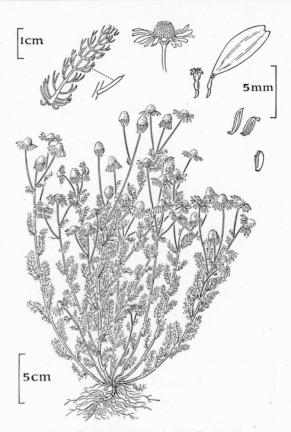

1362.   *Matricaria recutita* L.   Wild Chamomile
White and yellow

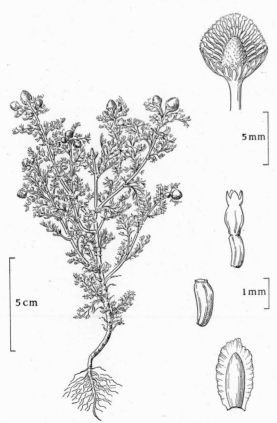

1363.   *Matricaria matricarioides* (Less.) Porter
Pineapple-Weed   Yellow

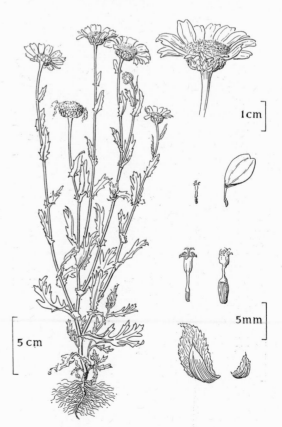

1364.   *Chrysanthemum segetum* L.   Corn Marigold
Yellow

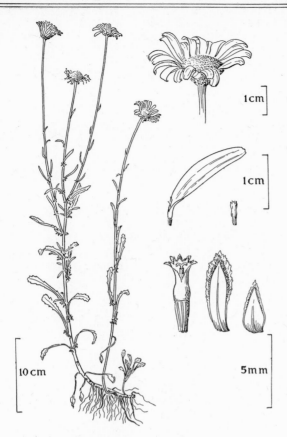

1365.   *Chrysanthemum leucanthemum* L.   **Ox-eye Daisy**
White and yellow

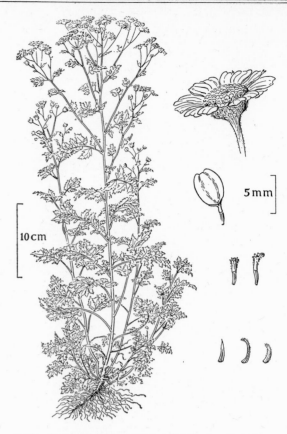

1366.   *Chrysanthemum parthenium* (L.) Bernh.   Feverfew
White and yellow

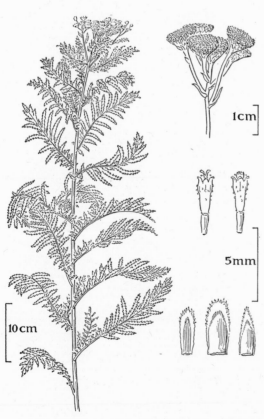

1367.   *Chrysanthemum vulgare* (L.) Bernh.   Tansy   Yellow

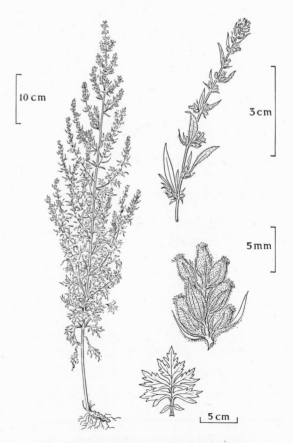

1368.   *Artemisia vulgaris* L.   Mugwort   Brownish

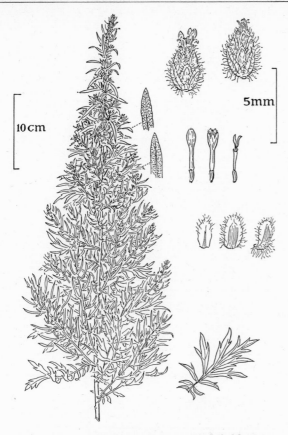

1369. *Artemisia verlotorum* Lamotte   Verlot's Mugwort
Brownish

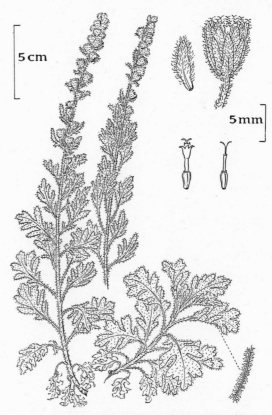

1370. *Artemisia stellerana* Bess.   Dusty Miller   Yellow

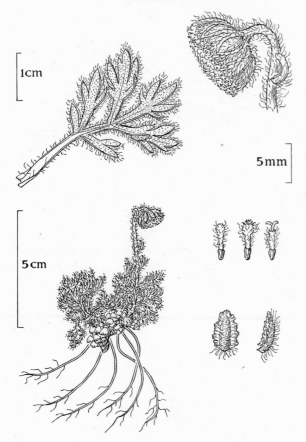

1371. *Artemisia norvegica* Fr.   Norwegian Mugwort
Yellowish

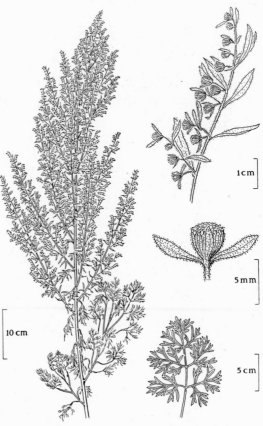

1372. *Artemisia absinthium* L.   Wormwood   Yellow

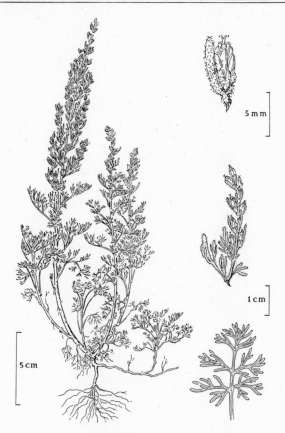

1373. *Artemisia maritima* L. Sea Wormwood
Yellowish or reddish

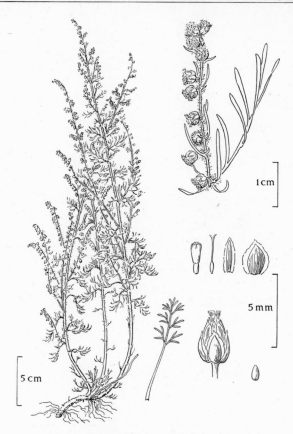

1374. *Artemisia campestris* L. Field Southernwood
Yellowish or reddish

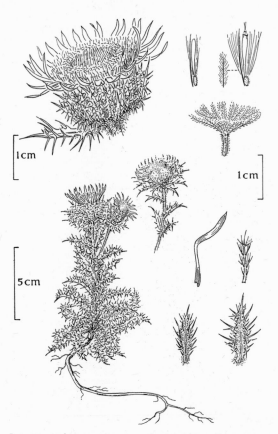

1375. *Carlina vulgaris* L. Carline Thistle Yellowish

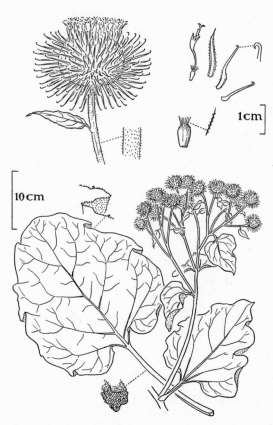

1376. *Arctium lappa* L. Great Burdock Reddish-purple

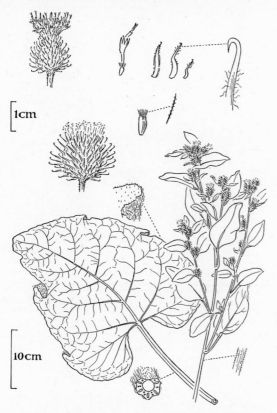

1377. *Arctium minus* Bernh.   Lesser Burdock   Reddish-purple

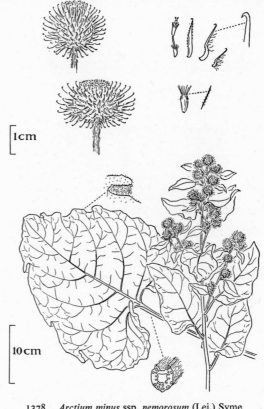

1378. *Arctium minus* ssp. *nemorosum* (Lej.) Syme
Reddish-purple

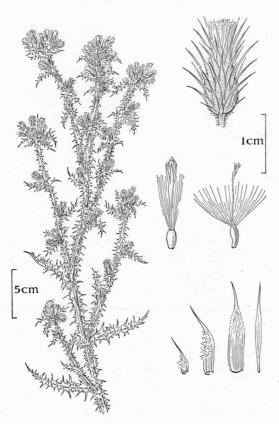

1379. *Carduus tenuiflorus* Curt.   Slender Thistle
Pale purplish

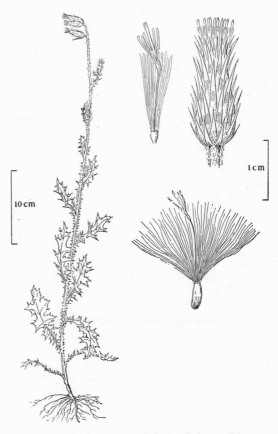

1380. *Carduus pycnocephalus* L.   Pale purplish

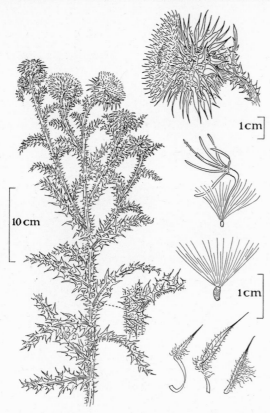

1381. *Carduus nutans* L.   Musk Thistle   Red-purple

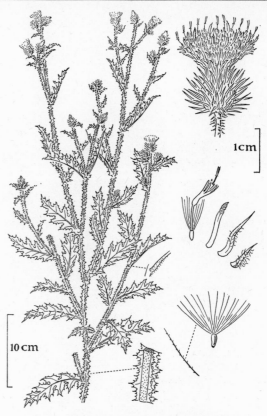

1382. *Carduus acanthoides* L.   Welted Thistle
Red-purple or white

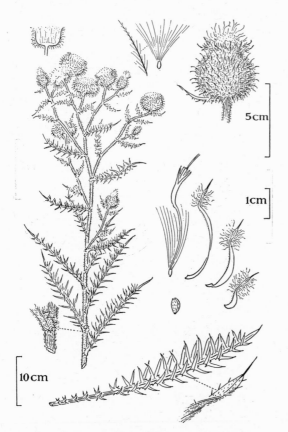

1383. *Cirsium eriophorum* (L.) Scop.   Woolly Thistle
Red-purple

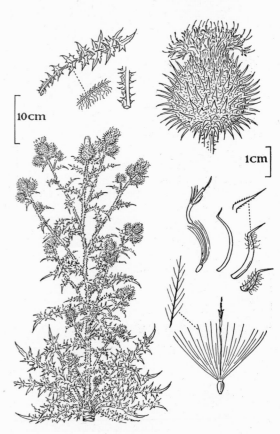

1384. *Cirsium vulgare* (Savi) Ten.   Spear Thistle
Red-purple

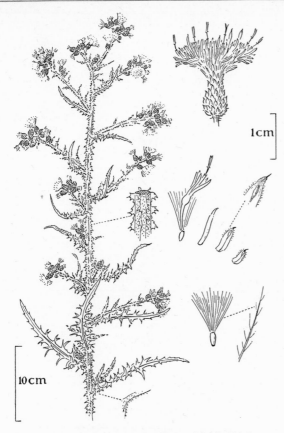

1385.   *Cirsium palustre* (L.) Scop.   Marsh Thistle
Red-purple

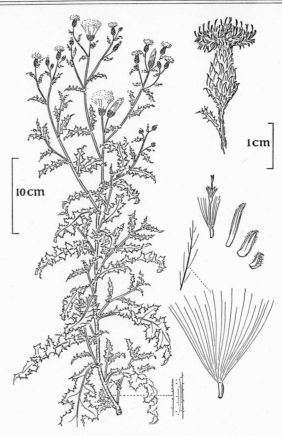

1386.   *Cirsium arvense* (L.) Scop.   Creeping Thistle
Pale purple or whitish

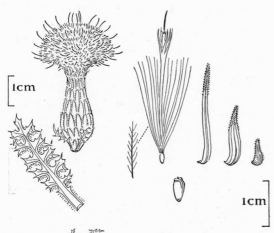

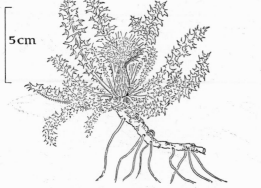

1387.   *Cirsium acaulon* (L.) Scop.   Stemless Thistle
Red-purple

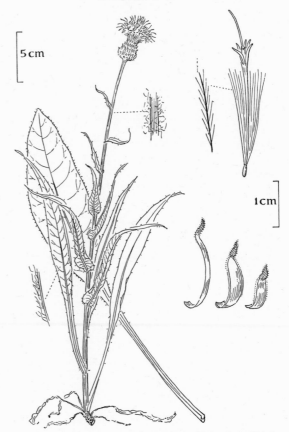

1388.   *Cirsium heterophyllum* (L.) Hill   Melancholy Thistle
Red-purple

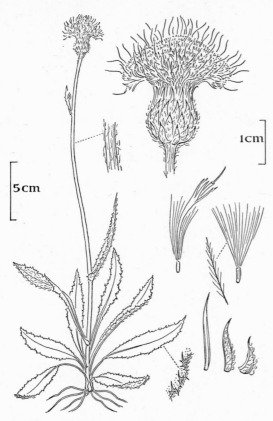

1389.   *Cirsium dissectum* (L.) Hill   Meadow Thistle
Red-purple

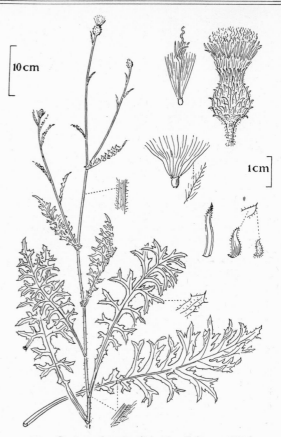

1390.   *Cirsium tuberosum* (L.) All.   Tuberous Thistle
Red-purple

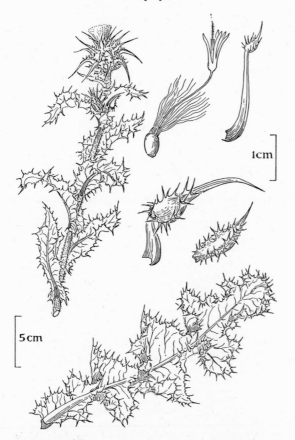

1391.   *Silybum marianum* (L.) Gaertn.   Milk-Thistle
Red-purple

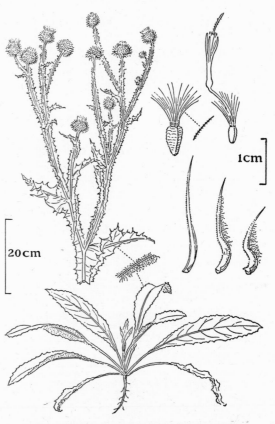

1392.   *Onopordum acanthium* L.   Scotch Thistle
Pale purple

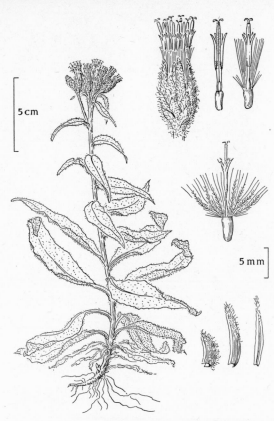

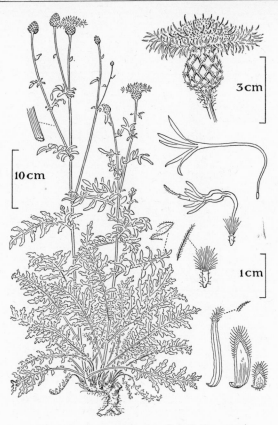

1393.   *Saussurea alpina* (L.) DC.   Alpine Saussurea
White and purple

1394.   *Centaurea scabiosa* L.   Greater Knapweed
Red-purple

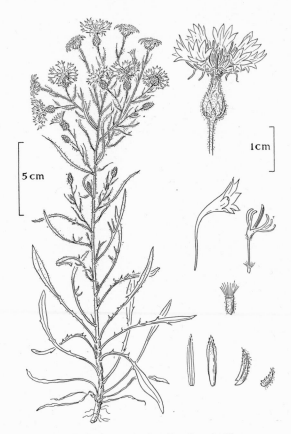

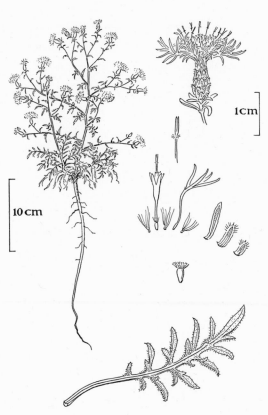

1395.   *Centaurea cyanus* L.   Cornflower   Blue

1396.   *Centaurea paniculata* L.   Panicled Knapweed
Purple

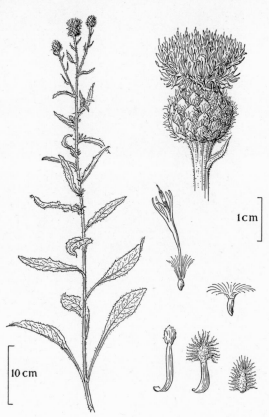

1397. *Centaurea nigra* L. Hardheads Red-purple

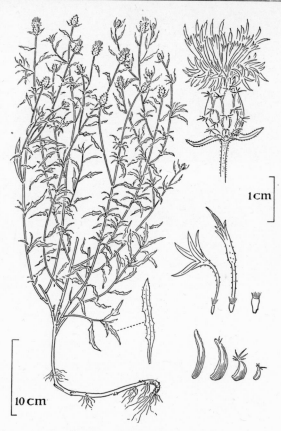

1398. *Centaurea aspera* L. Rough Star Thistle
Pale red-purple

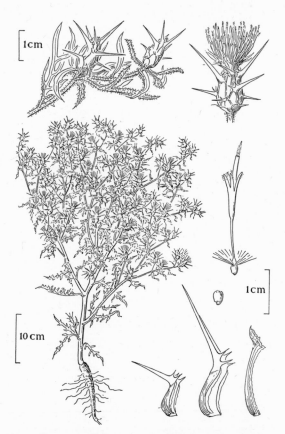

1399. *Centaurea calcitrapa* L. Star Thistle Pale red-purple

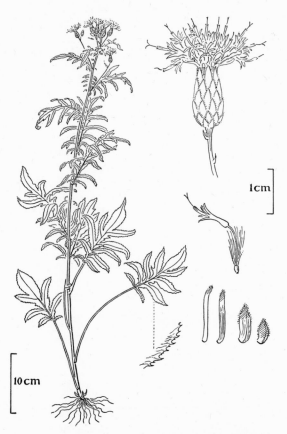

1400. *Serratula tinctoria* L. Saw-wort Red-purple

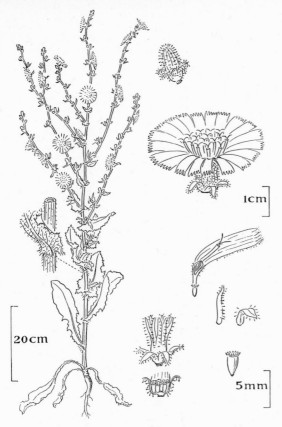

1401. *Cichorium intybus* L.   Chicory   Blue

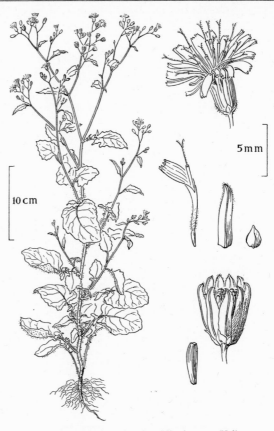

1402. *Lapsana communis* L.   Nipplewort   Yellow

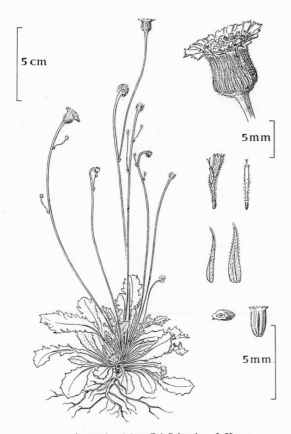

1403. *Arnoseris minima* (L.) Schweigg. & Koerte
          Lamb's Succory   Yellow

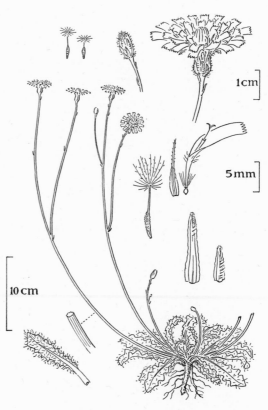

1404. *Hypocheris radicata* L.   Cat's Ear   Yellow

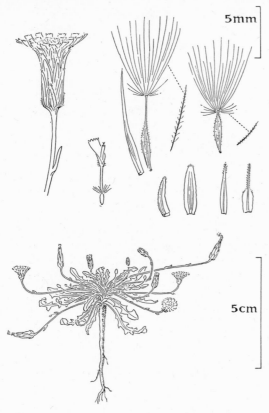

1405. *Hypocheris glabra* L. Smooth Cat's Ear Yellow

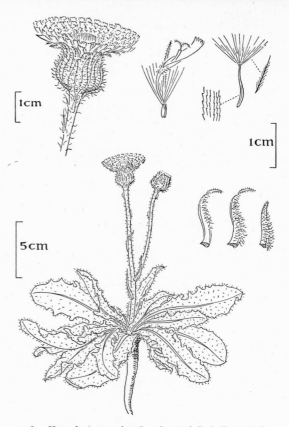

1406. *Hypocheris maculata* L. Spotted Cat's Ear Yellow

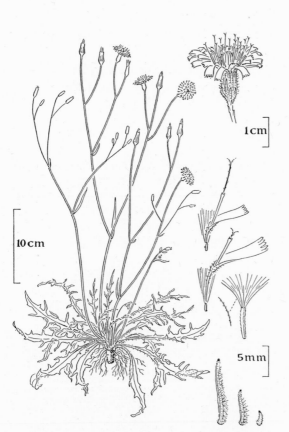

1407. *Leontodon autumnalis* L. Autumnal Hawkbit
Yellow

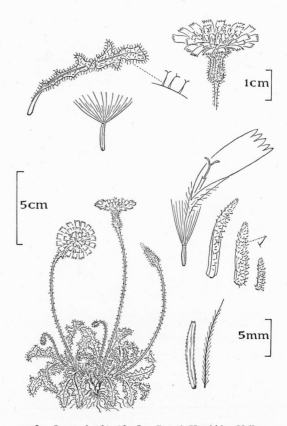

1408. *Leontodon hispidus* L. Rough Hawkbit Yellow

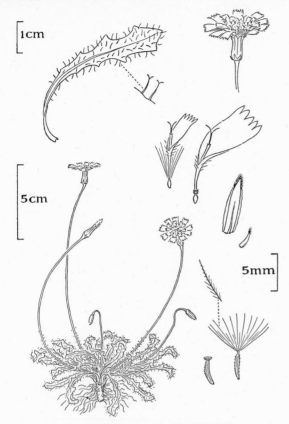

1409. *Leontodon taraxacoides* (Vill.) Mérat
Hairy Hawkbit   Yellow

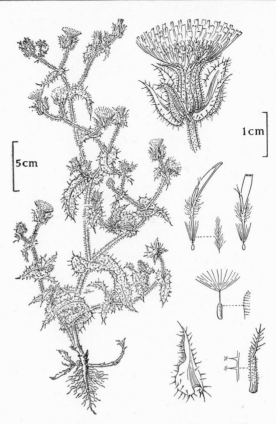

1410. *Picris echioides* L.   Bristly Ox-Tongue   Yellow

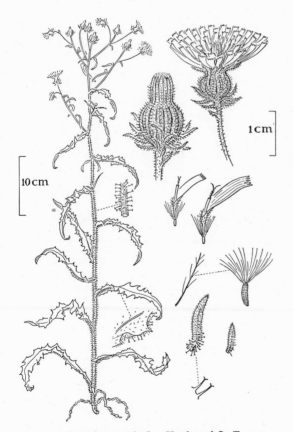

1411. *Picris hieracioides* L.   Hawkweed Ox-Tongue
Yellow

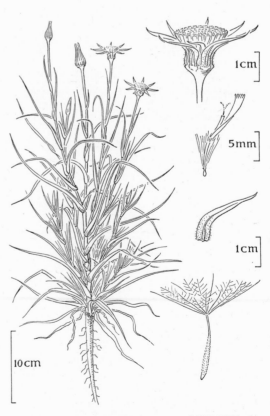

1412. *Tragopogon pratensis* ssp. *minor* (Mill.) Wahlenb.
Goat's-Beard   Yellow

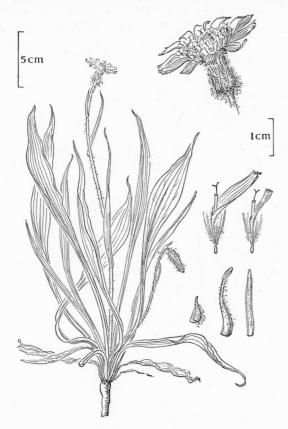

1413. *Scorzonera humilis* L.   Dwarf Scorzonera   Yellow

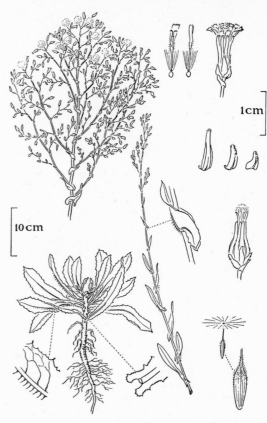

1414. *Lactuca serriola* L.   Prickly Lettuce   Yellow

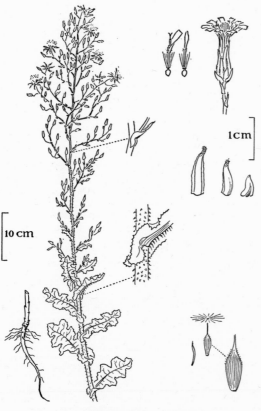

1415. *Lactuca virosa* L.   Yellow

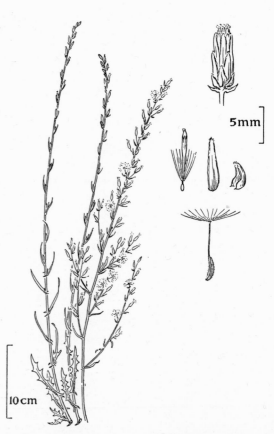

1416. *Lactuca saligna* L.   Least Lettuce   Yellow

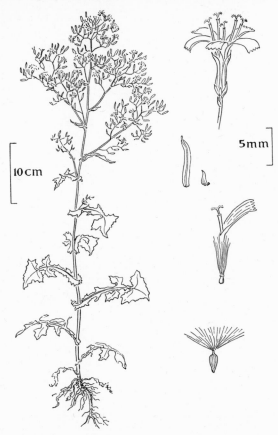

1417. *Mycelis muralis* (L.) Dum.   Wall Lettuce   Yellow

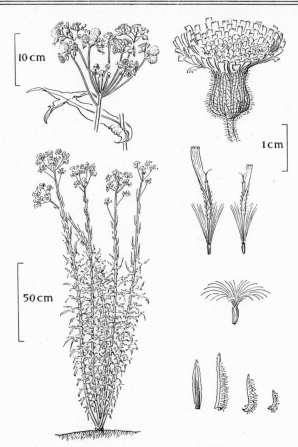

1418. *Sonchus palustris* L.   Marsh Sow-Thistle   Yellow

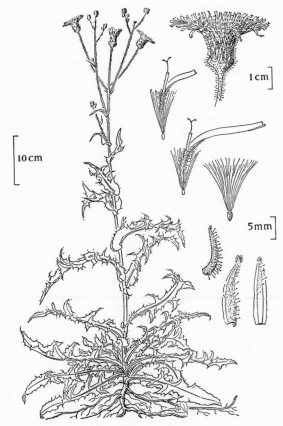

1419. *Sonchus arvensis* L.   Field Milk-Thistle   Yellow

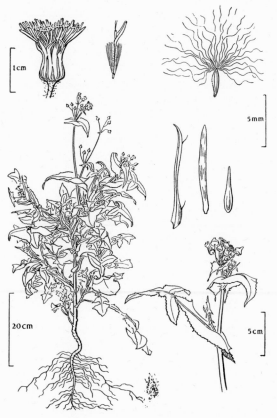

1420. *Sonchus oleraceus* L.   Sow-Thistle   Yellow

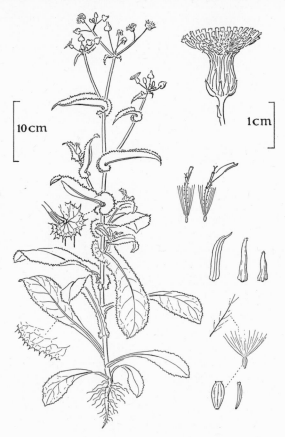

1421.   *Sonchus asper* (L.) Hill   Spiny Sow-Thistle   Yellow

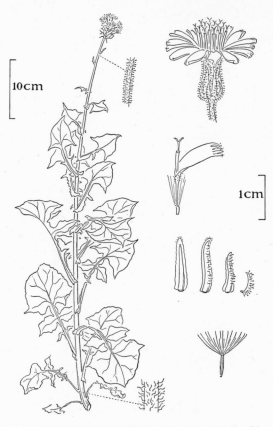

1422.   *Cicerbita alpina* (L.) Wallr.   Blue Sow-Thistle   Blue

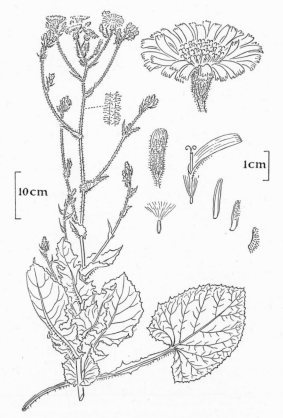

1423.   *Cicerbita macrophylla* (Willd.) Wallr.   Lilac

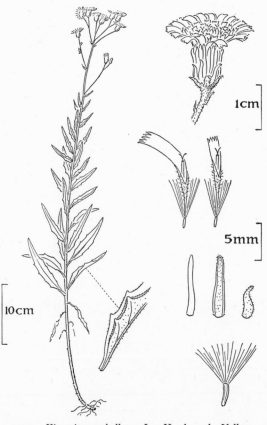

1424.   *Hieracium umbellatum* L.   Hawkweed   Yellow

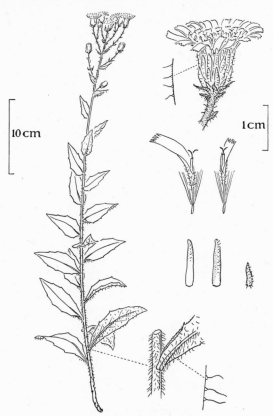

1425. *Hieracium perpropinquum* (Zahn) Druce   Hawkweed
Yellow

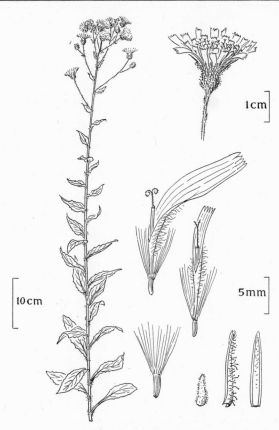

1426. *Hieracium latobrigorum* (Zahn) Roffey   Hawkweed
Yellow

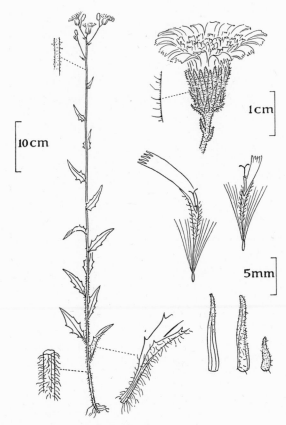

1427. *Hieracium tridentatum* Fr.   Hawkweed   Yellow

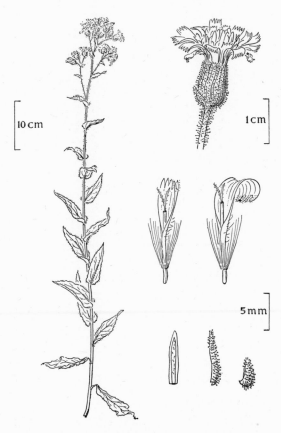

1428. *Hieracium prenanthoides* Vill.   Prenanth Hawkweed
Yellow

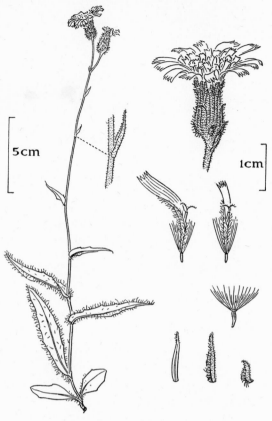

1429.   *Hieracium hethlandiae* (Hanb.) Pugsl.   Hawkweed
Yellow

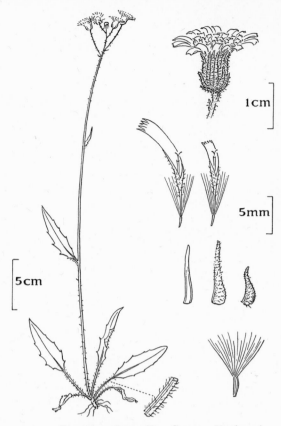

1430.   *Hieracium vulgatum* Fr.   Common Hawkweed
Yellow

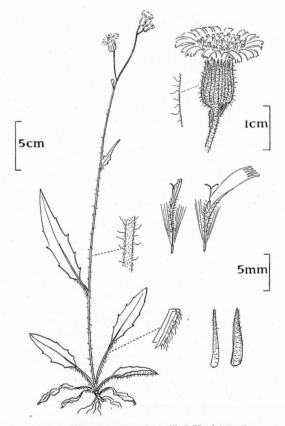

1431.   *Hieracium cravoniense* (F. J. Hanb.) Roffey
Hawkweed   Yellow

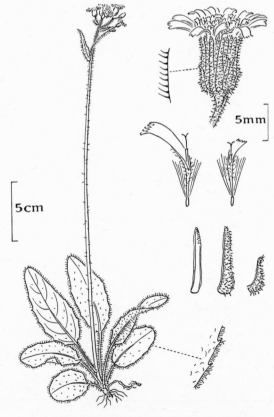

1432.   *Hieracium euprepes* F. J. Hanb.   Hawkweed
Yellow

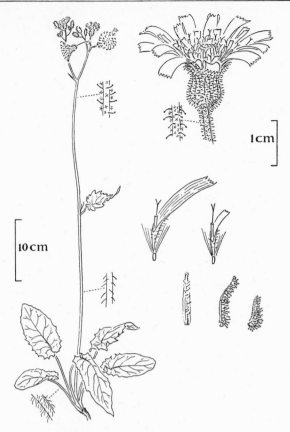

1433.   *Hieracium exotericum* Bor.   Hawkweed   Yellow

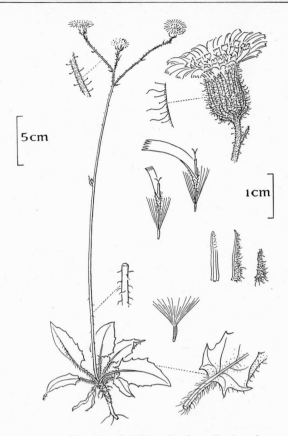

1434.   *Hieracium cymbifolium* Purchas   Hawkweed
Yellow

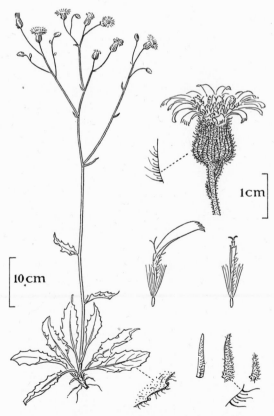

1435.   *Hieracium buglossoides* Arv.-Touv.   Hawkweed
Yellow

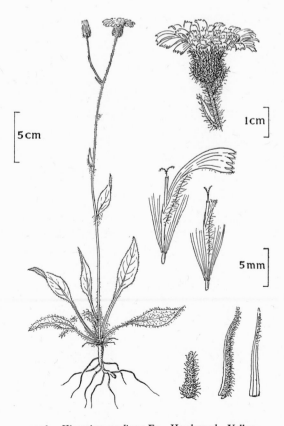

1436.   *Hieracium anglicum* Fr.   Hawkweed   Yellow

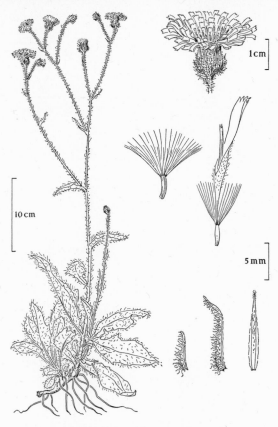

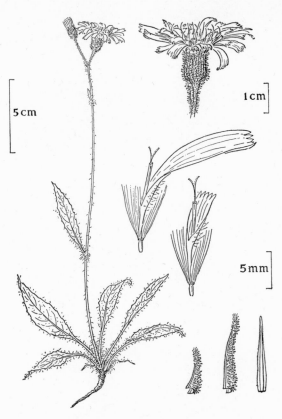

1437. *Hieracium iricum* Fr.   Hawkweed   Yellow

1438. *Hieracium lingulatum* Backh.   Hawkweed   Yellow

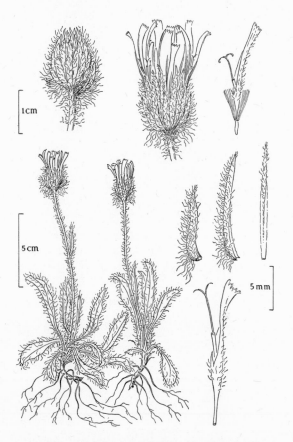

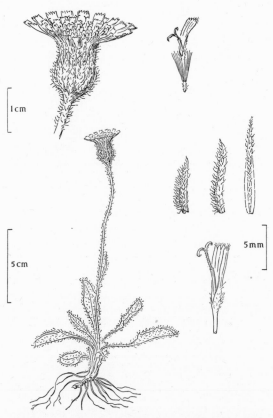

1439. *Hieracium holosericeum* Backh.   Hawkweed   Yellow

1440. *Hieracium nigrescens* Willd.   Hawkweed   Yellow

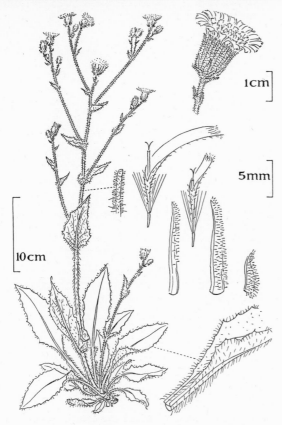

1441. *Hieracium speluncarum* Arv.-Touv. Hawkweed
Yellow

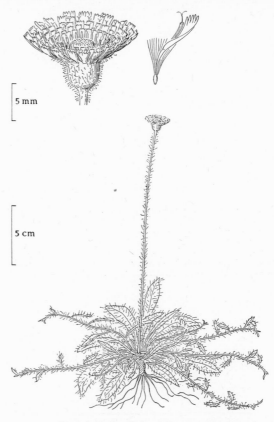

1442. *Hieracium pilosella* L. Mouse-ear Hawkweed
Yellow

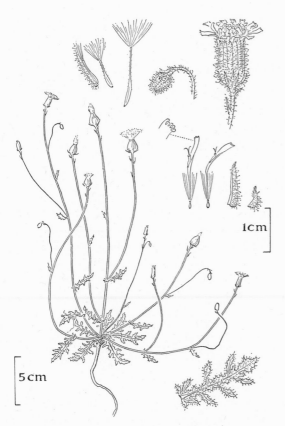

1443. *Crepis foetida* L. Stinking Hawk's-beard Yellow

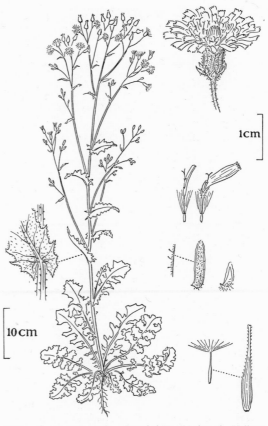

1444. *Crepis vesicaria* L. Bearded Hawk's-beard Yellow

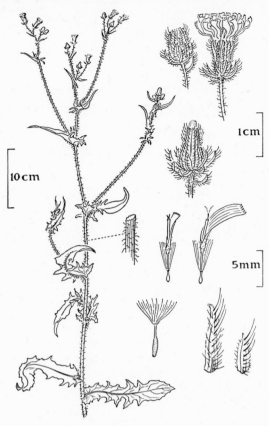

1445. *Crepis setosa* Haller f. Bristly Hawk's-beard
Yellow

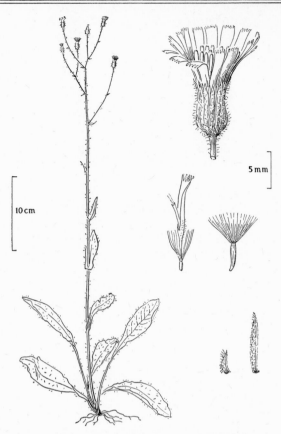

1446. *Crepis mollis* (Jacq.) Aschers. Soft Hawk's-beard
Yellow

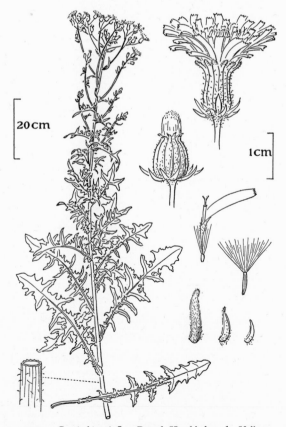

1447. *Crepis biennis* L. Rough Hawk's-beard Yellow

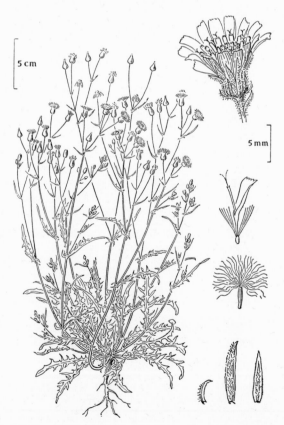

1448. *Crepis capillaris* (L.) Wallr. Smooth Hawk's-beard
Yellow

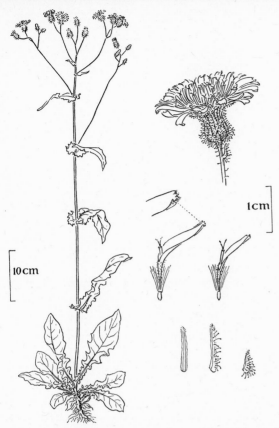

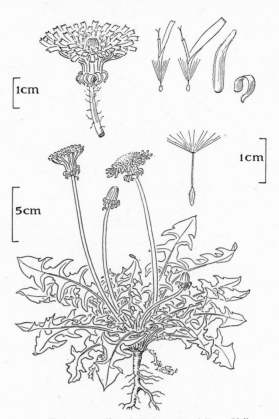

1449.   *Crepis paludosa* (L.) Moench   Marsh Hawk's-beard
               Yellow

1450.   *Taraxacum officinale* Weber   Dandelion   Yellow

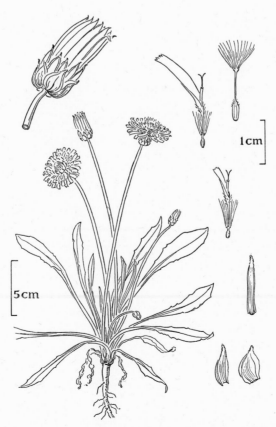

1451.   *Taraxacum palustre* (Lyons) DC.   Dandelion   Yellow

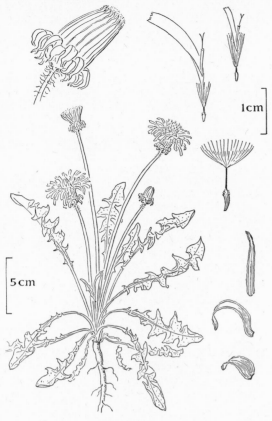

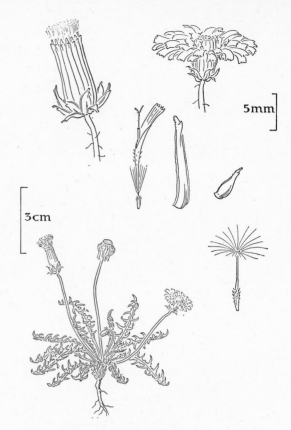

1452.   *Taraxacum spectabile* Dahlst.   Dandelion   Yellow

1453.   *Taraxacum laevigatum* (Willd.) DC.   Dandelion
Yellow

INDEX

[113]